Oxford Applied Mathematics and Computing Science Series

OXFORD APPLIED MATHEMATICS AND COMPUTING SCIENCE SERIES

I. Anderson: *A First Course in Combinatorial Mathematics*

D. W. Jordan and P. Smith: *Nonlinear Ordinary Differential Equations (Second Edition)*

D. S. Jones: *Elementary Information Theory*

B. Carré: *Graphs and Networks*

A. J. Davies: *The Finite Element Method*

W. E. Williams: *Partial Differential Equations*

R. G. Garside: *The Architecture of Digital Computers*

J. C. Newby: *Mathematics for the Biological Sciences*

G. D. Smith: *Numerical Solution of Partial Differential Equations (Third Edition)*

J. R. Ullmann: *A Pascal Database Book*

S. Barnett and R. G. Cameron: *Introduction to Mathematical Control Theory (Second Edition)*

A. B. Tayler: *Mathematical Models in Applied Mechanics*

R. Hill: *A First Course in Coding Theory*

P. Baxandall and H. Liebeck: *Vector Calculus*

D. C. Ince: *An Introduction to Discrete Mathematics and Formal System Specification*

P. Thomas, H. Robinson, and J. Emms: *Abstract Data Types: Their Specification, Representation, and Use*

D. R. Bland: *Wave Theory and Applications*

R. P. Whittington: *Database Systems Engineering*

J. J. Modi: *Parallel Algorithms and Matrix Computation*

P. Gibbins: *Logic with Prolog*

PETER GIBBINS

Open University

Logic with Prolog

CLARENDON PRESS · OXFORD
1988

Oxford University Press, Walton Street, Oxford OX2 6DP

Oxford New York Toronto
Delhi Bombay Calcutta Madras Karachi
Petaling Jaya Singapore Hong Kong Tokyo
Nairobi Dar es Salaam Cape Town
Melbourne Auckland

and associated companies in
Berlin Ibadan

Oxford is a trade mark of Oxford University Press

Published in the United States
by Oxford University Press, New York

British Library Cataloguing in Publication Data

Gibbins, Peter
Logic with Prolog.—(Oxford applied
mathematics and computing science series).
1. Computer systems. Programming languages:
Prolog
I. Title II. Series
005.13'3

ISBN 0–19–859671–5
ISBN 0–19–859659–6 (Pbk.)

Library of Congress Cataloging in Publication Data

Gibbins, Peter.
Logic with prolog.
Bibliography: p.
Includes index.
1. Prolog (Computer program language) 2. Logic
programming. I. Title.
QA76.73.P76G53 1988 005.13'3 88–25287

ISBN 0 19 859671 5
ISBN 0 19 859659 6 (Pbk.)

Printed in Great Britain
at the University Printing House, Oxford
by David Stanford
Printer to the University

To
Mark Weaver

Preface

One of the many lessons to be learned from the history of science and technology is that the apparently least practical and purest research of today often becomes the applied science of tomorrow, and the basis for new technology. So it was with formal logic.

Formal logic began as the study of valid everyday human reasoning, a weapon for philosophers. Logic enabled one philosopher to defeat another in argument, and this was, for centuries, its only technological application. After more than two millennia in which progress in logic was very nearly indiscernible, formal logic severed its connection with everyday argument, and became first a tool for metaphysicians, and then a branch of pure mathematics. Now, through logic programming, formal logic is becoming worldly again. With Prolog we *program in logic*. Logic has become a programming language, a tool for technologists.

In this book we use Prolog program logical ideas. We prototype parsers, pretty-printers, interactive proof-checkers, programming language interpreters, and theorem-provers of various kinds. We implement our own versions of Prolog, in Prolog.

Prolog is a first shot at a logic programming language. Prolog has a peculiar logic of its own, and to understand Prolog and its peculiar logic we need to understand some conventional, unpeculiar, formal logic. A study of Prolog is therefore no substitute for a study of logic. But Prolog is a good medium in which to implement logic, and one gains a better understanding of logic, and of anything, if one tries to implement aspects of it in a programming language.

The book begins with a few chapters on Prolog as a programming language, and how one can use it. Chapter 2 contains a summary of most of the Prolog we need, together with examples which show how the

Prolog interpreter executes queries and programs.[1]

The core of the book, chapters 5 through 13, deals with the propositional and predicate calculi. The latter, enriched with identity, is the basic logic of mathematics and so of computing. It is called *first-order logic*. We deal with these systems conventionally, via natural deduction systems. We sketch the theory behind resolution theorem-proving. In the last two chapters we examine the extended logic of a small programming imperative language, and the restricted logic of real Prolog. In dealing with all these subjects we are less interested in theoretical rigour — there are no completeness proofs — and more interested in ideas and in how they may be programmed in Prolog.

Programming logic is one way in which we can increase our understanding of logic. But we also consider *philosophical* questions — questions that form part of the subject matter of the philosophy of computing science. How should we view Prolog? Should we really view Prolog as an implementation of logic, as some of its protagonists seems to argue? Or should we view it merely as a high-level procedural programming language ? Might it be that Prolog will change our ideas of what logic is, as some computing scientists suggest? Might it be that logic programming will revolutionize software engineering?

The book contains a good deal of Prolog code. All the Prolog code in the book is *C-prolog*.It was written on a variety of VAXes running either VMS or Unix™, and then taken down to an Apple Macintosh™. The book itself was produced on a Macintosh™, using Writenow™, printed on an Apple Laserwriter™, and reduced to 67% for final production. Code taken down from a VAX running C-Prolog appears in **bold**. Sessions photo-logged as they appeared on the terminal screen are reproduced again in **bold**, with the queries supplied by the user written in *italic*.

Chapters 9 and 13 contain quite a number of natural deduction proofs. These were produced not on a VAX but on an ancient BBC 'B' machine running a proof-checker written in BBC Basic and 6502 machine code. The proof-checker is commercially available[2]. The checked proofs were given minimal editing.

---

[1] Chapter 2 is not meant to be a substitute for a full account of the language and its use, such as you can find in *Programming in Prolog* by WF Clocksin and CS Mellish, *The Art of Prolog* by Leon Sterling and Ehud Shapiro, MIT Press, 1986, or *Prolog: A Relational Language and Its Applications* by John Malpas, Prentice-Hall International, 1987.

[2] From Routledge and Kegan Paul under the title *Logic+*, PF Gibbins, 1985. The programs in *Logic+* were written to support Bill Newton-Smith's *Logic: A First Course*, Routledge and Kegan Paul,1985.

Parts of this book we given in two lecture courses in the Department of Computer Science at Bristol University during 1985 and 1986. One course covered discrete mathematics and logic in computing, and was aimed at undergraduates. The other was a master's degree course on the principles of programming languages. I should like to thank my students there, especially David Andrews among the undergraduates, for their stringent criticism.

In addition, I am pleased to record my debt to my friend Bill Newton-Smith of Balliol College, Oxford. The forms that the natural deduction systems for propositional and predicate logic take in this book were Bill's creations, and my version of first-order logic is only a modest extension of the predicate logic of the one in Bill's *Logic: A First Course*. It was Bill who encouraged me to write the book in the first place.

I should also like to thank Ray Weedon of the Open University, and Tony Dale of the University of Hull, both of whom read much of the text, and made detailed and helpful comments.

The Open University P. G.
April 1988

Contents

1 Why logic matters to computing science

We begin with two questions:

what is computing science?

why does logic matter to computing science?

Notice that we say *computing* science and not *computer* science. We are signalling the fact that in this book we are not interested in hardware, in real physical machines. We are interested in the scope of the *activity* called computing.

An abstract answer to the first question is: computing science is about the computable. Computing science deals with *representations* of the computable, and with the relationships that obtain between representations of the computable. More precisely, computing science is about computable mathematical functions, and the representations we make of them. We, as programmers, tend to think of machine code as the fundamental representation of our computation, the one which really runs on the machine. But this is because we, as programmers, tend to lose interest in representations at lower levels than machine code, which is therefore the lowest level at which we are likely to find ourselves programming a machine. But there is nothing absolutely fundamental about machine code. Some processors can be programmed at the even lower level of micro-code. A program written in the programming language Pascal, and the same program written in the machine code to which it compiles, are two different representations of the same computable function, the relationship between them will be effected by a compiler, an object worthy of study by the computing scientist and one which may itself have been written in Pascal, or machine code.

So computing science is partly about representability and representation. It is partly about what can represented in the form of a computer program, and about the *levels* at which such representations can be made. Software engineering, which is in part simply applied computing science, concerns itself with managing the process of representation, the process which takes us from a description of what software is to do — its specification — to its representation as source code in a real programming language.

What mathematics does one need to know to study representations of the computable and the relationships between them? What mathematics does the computer scientist and the software engineer need to know? The answer is logic and discrete mathematics. A theoretical physicist must know calculus, and much more. An engineer — civil, aeronautical, mechanical, electronic engineer — cannot afford to remain entirely ignorant of calculus. Calculus gives much of physics its mathematical basis and it also provides the engineer with the apparatus with which to apply the theories of physics to the business of making things work in the everyday world. Similarly, logic and discrete mathematics form the mathematical basis of theoretical computing science. Why logic, and why discrete mathematics? Take discrete mathematics first.

Physics deals with quantities which vary continuously, like position in space and time, like mass, and momentum. Computing science deals with quantities which do not vary continuously, but which are discrete — like whether something is or is not the case, whether a device is 'on' or 'off', whether a voltage is 'high' or 'low'. The mathematics of the continuum is by and large quite irrelevant to computing science, which is one reason why computing is quite different from physics. Like the mechanical engineer, the software engineer needs to know some mathematics, and her mathematics is discrete mathematics and logic. It really would be surprising if the mathematics of software engineering and the mathematics of computing science were very different sorts of mathematics. For the mathematics of software engineering is just the applied mathematics of theoretical computing science.

'Software engineering' is more than just one of those fashionable phrases of which the discipline of computing is so full. It has the merit of carrying an important hidden message — that writing, designing, specifying computer programs, representations of the computable, must be treated as a form of engineering, an activity subject to the demand that its products have a guaranteed quality. To guarantee that a piece of software is of the required quality demands that we have methods for developing software of high quality. It also demands that we can show that our product has the required quality.

A piece of software is a strange engineering product, compared with those engineering products we have become used to. It isn't visible like an aeroplane, though like an aeroplane it can crash (and more than once). Software is a mathematical product. We test an aeroplane by subjecting it to stresses in a controlled environment, and not by throwing passengers into it and sending it on its way. We can't prove that it is airworthy. But software is a mathematical product, not a physical artefact[1]. We should be able to do more than 'test' a piece of software. We ought to be able to *prove* that it is crash-proof, if that is what we demand of it. This is one place where logic gets in to software engineering, via the business of proving that the software we develop is correct.

1.1 Software engineering and logic programming

First we develop the connection between logic and software engineering. On one side, computing science interfaces with software engineering, as physics interfaces with and feeds into mechanical and electronic engineering. The mathematics of the corresponding pure science and that of its applied technology are similar. The relation between computing science and software engineering is even closer. It is no exaggeration to say that the technology of software is the technology of logic and discrete mathematics. Once we study the capacities, practical and theoretical, of the idealized computer — as we do in computing science — we naturally go on to develop efficient methods of

[1] There are analogies and some disanalogies between physics and computing science. Unlike other sciences, computing science is not a study of naturally occurring objects or events or processes. Physics mostly studies atoms, the solid state, the electromagnetic field, space-time and so on. All these things exist independently of human beings, though human beings have discovered them. Computing science, narrowly conceived, is the study of the information-processing machines that we have *invented*, and can build, and can *envisage* building. It is the study of all possible computing machines, of their computational powers, both practical and theoretical, and of how they may be formally described and controlled.

Should we think of computing science as a science at all? The answer is that we can, because there are *objective* limits to and constraints upon the powers that any conceivable computer might attain. We may invent computers, but the physical world and the world of mathematics restrict the range of the possible. We are, of course, stuck with the physical world. We might expect the speed of light and Planck's constant to impede our progress towards the omnipotent computer. But the range of what computable is in any case limited, as we shall see. Establishing what these limits are is part of the subject matter of computability theory, an offshoot of logic. Here again we find logic entering computing science, at a very fundamental level indeed.

controlling computers.

If computing science studies what computing machines may be and how they may be described, software engineering studies how the behaviour of computing machines may be prescribed. Software engineering deals with the business of producing software, all the way from its specification to implementation. But what, in more detail, is the aim of software engineering, and how should we go about achieving it?

In computing, complexity is costly, and is liable to generate errors. Simplicity and cheapness tend to go hand in hand. But simplicity isn't an absolute. In science one usually buys simplicity through mathematics, which has its own complexity. Physics gives us an elegant description of the world only because it succeeds in mathematizing Nature. It is the same in software engineering. When we produce programs, we want to minimize our intellectual burden and hence the cost to us, but we can do this best by making an initial investment in the appropriate tools. The appropriate tools of the software engineer include formal logic, as we shall see in more detail in later chapters.

Here is a first answer to the question of why logic matters to computing science, the answer of the software engineer. Logic — formal or mathematical logic — matters to computing science because computers run programs and because programs have a logical structure. Logic forms a sizeable chunk of the language we use to specify the behaviour of the software we mean to write. The programs we do write are either correct or incorrect, with respect to their specifications. A logic of correctness is a tool for showing that a program is correct (if it is).

Logic supplies the machinery which enables us to justify the programs we write, and so enables us to increase our confidence in our software. Therefore, formal logic is an indispensable part of the software engineer's armoury. To repeat, software engineering demands that mathematical logic be put to new use as an applied, technological mathematics. Logic matters to computing science first because it is a large part of the applied mathematics of software engineering.

But logic matters to computing science secondly, because computers can solve problems, often in ways that seem to us to exhibit a logical intelligence. Computers can solve problems which seem to require knowledge. Computers can be made to play chess quite well, and even to prove theorems in mathematical logic. Formal logic describes structures which enable us to represent knowledge. Logic provides the structures for representing propositional knowledge, knowledge expressible in words. It also describes and prescribes the interrelationships which hold between facts. One fact can entail another and logic can tell us when and why. And so logic provides the inferential mechanism for an important class of knowledge-based systems, in which there is now so much interest among workers in artificial intelligence (AI, for short).

Logic impinges on computing science through artificial intelligence and through software engineering. In recent years, these two apparently quite disparate computing disciplines have threatened to converge. AI on the one hand and software engineering on the other, may seem to have little in common. However, the results of work in AI are beginning to be felt even in the real world of practical program building, the world that interests the software engineer.

Nowadays, computers can be programmed to make logical inferences, an activity which consumed much of early work in AI. For if human thinking and reasoning involve making inferences, some parts of artificial intelligence must presumably involve the programming of logical inference. There is a new style of programming — logic programming — which, having begun as a branch of AI, now consists in part in making general purpose programs look as much as possible like statements in formal logic.

Logic programming involves the use of logic programming languages — Prolog is the best known — which enable one to write programs that approximate to a collection of purely logical statements. Logic programming promises a quantum leap both in the elegance and clarity of software and in the productivity of programmers.

How can logic programming help the software engineer? A logic program can — in theory at least — omit all mention of the flow of control on which the software designer is usually forced to focus her effort. Relieved of the burden of designing the flow of control in her programs the software designer can concentrate on the logic of the program. Logic programs are shorter than and, again in theory, can be more easily seen to be correct than programs written in conventional programming languages. And yet a logic programming language like Prolog has no less power, in principle, than a conventional imperative language like FORTRAN, Pascal, or ADA.

Now we have two reasons for investing effort in the study of logic, even as software engineers, let alone as theoretical computing scientists. Logic is a vital support in conventional programming methodology through the power it gives us to prove a program correct with respect to its specification. And logic is the basis for the new field of logic programming, which developed out of research in artificial intelligence and which is beginning to make its impact on conventional software engineering. But in saying both of these things we are only scratching the surface of the subject matter of computing science. For computing science is an essentially mathematical subject, and a part of its mathematics is mathematical logic.

The theory of computability, the theory of what computer programs mean, theories of formal languages, theories of data structures are all essentially mathematical and draw on the results of mathematical logic. And so logic matters to computing science thirdly, because it provides the subject with its mathematical foundation, just as calculus provides physics with its mathematical foundation. But in this book we mostly avoid deep questions about the limits of the computable. We mostly address the practical question: of what use is formal logic to the software engineer?

1.2 Logic in software engineering

A *software crisis* has been with us in name, since 1968,[2] and in reality, ever since programming in high level languages like FORTRAN and COBOL began. The roots of the software crisis lie in the growing power of computer hardware and the growth in the demands that have been made of the software that controls the new powerful machines. Notoriously, hardware costs spiral down by an order of magnitude every few years. The software engineer is overwhelmed by the sheer success and the power of the new, cheaper hardware. To her embarrassment, but possible profit, the cost of software takes an ever-increasing proportion of the overall cost of a computer system. Software is both more expensive, and much less reliable, than the hardware it runs on. As computers get more and more powerful so we need a better understanding of the software that controls them.

Software engineering begins with the problems: how can we manage the complexity of software systems? what methods can we use to enable us to specify and implement reliable and cheap software products?

Software lies at the interface between man and machine. Software is written by human beings and is language-like. Software has a logic because we must understand what we write. We should be able to treat a program as a mathematical text. A program should have a visible

[2] For an excellent collection of article written around the idea of a software crisis see *Programming Methodology*, David Gries (ed), Springer-Verlag, 1978.

underlying logic which ought to tell us whether or not it does what we want it to do. So why do we find such difficulty in writing software that is free of 'bugs'?

First of all, programs are not proofs of anything. Part of the difficulty of building software consists in the fact that programs written in conventional programming languages like Pascal are not like the pieces of mathematics we are familiar with. They are not valid or invalid. They are not correct or incorrect, except with respect to a specification. Establishing that a program behaves in accordance with its specification is usually treated as something quite separate from (and much more demanding than) writing the program.

Secondly, the programming languages in which they are written are more complex than the languages of ordinary mathematics. For example, take Pascal, one of the most elegant and best-designed programming languages in general use. Pascal has a variety of commands, data-types, facilities for defining new types, procedures, declarations, restrictions on the order in which declarations may be made. Programming languages always make some concession to the architecture of the machines on which they are implemented. The way we represent software is determined by a compromise between what computing machines can be made to do and what a human being can understand. Thus Pascal has the assignment statement because it is run on computers which easily implement the assignment statement. It has user-defined types and procedures so that the programmer can structure her program.

If we are to demonstrate the reliability of a program, we must find a way of reasoning about its text, about source-code, so that we can prove its correctness (with respect to a specification). It is no good relying on 'testing' a program with what we think of as appropriate test data. It is no good relying on the method throwing data at the program and 'de-bugging' it. As Edsgar Dijkstra said in one of the most frequently quoted of all the software engineer's golden rules:

Program testing can be used to show the presence of bugs, but never to show their absence ! [3]

Correctness means, among other things, absence of bugs. The

[3] The most over-quoted quotation in software engineering, but still excellent value. See Dijkstra's essay *Notes on Structured Programming*, p.6 in *Structured Programming*, O-J Dahl, EW Dijkstra, and CAR Hoare (eds), Academic, 1972.

concern for program correctness is now at least as great as the concern for efficiency in execution. Why? Because computing machines are now so powerful, so fast, have so much memory, and the programs that we demand to be run on them are now so complex that there has been a shift in our perception of the fundamental problem of programming. The software engineer needs to employ a programming methodology, a set of intellectual tools which enable him to manage the complexity of the specification, design and implementation of software.

A program is, as we said, correct only with respect to a specification. We must develop methods of specifying programs, and methods of proving them correct with respect to their specifications. The first problem leads the software engineer to develop formal specification languages. Experience teaches us that informal specification has all the disadvantages of informal reasoning. Informal specification is vague, and does not provide machinery for rigorous proofs of what a specification demands. The second problem leads the logician to develop logics of program correctness, formal systems of reasoning about programs as mathematical texts. All this presumes that there is a great gulf between specification and implementation, as there generally is in conventional programming in languages like Pascal. But in logic programming we try to close the gap.

So the software crisis takes us back to logic programming. Many computing scientists would say that the software crisis is so profound that we must narrow the gap between software and its logic to vanishing point so that software just is logic. Logic programming is just one of the possible solutions to the problem of engineering cheap, reliable software, the problem of managing the complexity of software, a solution to which consists in paring down the process of writing programs to the point where it consists in no more than expressing the logical structure of the data and the program that handles it.

Put briefly, current software engineering presumes that conventional programming is done better when its logic is used as a guide; that we need to develop specification languages based on logic and discrete mathematics; that we must develop methods for implementing specifications written in specification languages and showing them to be correct; that all this requires logic, indeed new logics of program correctness.

On the other hand, the proponents of logic programming are more radical. They suppose that we take as an ideal the 'equation'

Programs = Logic + Control. [4]

Logic programmers demand that the *Control* on the right-hand side of the equation be taken over by the programming language, and not be left to the programmer.

Here we have two competing ideologies in software engineering. There is a dispute between those who want to manage with imperative languages like Pascal and those who do not. Those who don't often favour logic programming over conventional software engineering. There is even a compromise position. One can use a logic programming language like Prolog to *prototype* software which will eventually be implemented in a conventional, imperative programming language like Pascal, or ADA. Prolog programs are relatively short, as we shall see. And the rapidity of *rapid prototyping* in Prolog can help us learn and test our specifications, before we produce the final implementation of our system. Whatever the outcome of these ideological battles within software engineering, most contemporary computer scientists and software engineers would say that a study of logic is a more appropriate requirement for computer scientists than the more usual study of calculus.[5] Yet hitherto, logic has been strangely and unduly neglected in computing science. Its time has clearly come. We can expect to see formal logic make an enormous impact on computing in the next few years and, in turn, we can expect to see computing science make changes in the subject matter of formal logic.

1.3 From syllogism to algorithm

But there are also deeper, perhaps less immediately practical reasons for the computing scientist to interest himself in formal logic, than those thrown up by the discipline of software engineering and the project of developing artificial intelligence. Computing science itself has its roots in two quite different disciplines, mathematics and electrical engineering, one highly abstract and apparently pure, and the other very

[4] See *Introduction to Logic Programming,* Hogger, Academic Press, 1984, p.99ff but also Kowalski *Logic for Problem Solving* p.125ff. With today's Prolog one reasonably claim that *Programs = Logic + Control* and that Prolog handles *some*, though not all, aspects of *Control* already, as we shall see.

[5] As Manna and Waldinger say in their *Logical Basis for Computer Programming,* Addison-Wesley, p. vii, logic 'should replace calculus as a requirement for undergraduate majors'.

concrete and practical. On the one hand, the study of what can and what cannot be computed grew out of the new mathematical logic of the late nineteenth and early twentieth centuries. This new logic was itself a response to abstract philosophical puzzles in the foundations of mathematics. The theoretical limitations on the computing power of all conceivable computers are of interest to the mathematician. On the other hand, men who knew more about electronics than about mathematics began to build primitive electronic computing machines in the 1930s. And thus began a branch of electronic technology whose growth is still explosive and whose products constantly increase in power. But the early mathematical heroes of computing science were not engineers but were logicians, like Alan Turing and John von Neumann, which makes it very odd that formal logic has until recently been neglected in the education of computer programmers. Convinced that logic is important to computing science we should now ask what it, logic, really is. We can approach it in two ways, one 'pure' and historical, the other 'applied' and a matter of engineering.

Logic, traditionally conceived, is the study of valid reasoning, and so logic has a much longer history than computing. To understand how logic developed in the way that it did, how it came to be considered to be the foundation of mathematics and of computing, and how it may be useful to the computer scientist, you have to go back to the fourth century BC on the one excursion into philosophy which every computing scientist should enjoy, if she enjoys no other. Formal logic is a human discovery, or invention, of genius. We owe it to Aristotle who, in the fourth century BC, set about codifying the forms of valid reasoning that could be used in informal argument and in mathematics. One of Aristotle's motives in constructing his logic was to sort out the valid arguments from the invalid arguments pedalled by the Sophists, the professional philosophers of his day. This philosophical motive persists in logic to the present day, though it is often pursued in conjunction with another related goal, one which is relevant to the computing discipline of artificial intelligence, namely the fact that logic can help us understand the structure of natural languages like Greek, and English. This is one of the routes by which logic gets into artificial intelligence, via the problem of knowledge representation. Aristotle's logic, as it has come down to us, is almost wholly restricted to the working out of syllogisms, arguments of the

All men are mortal,
Socrates is a man,
therefore Socrates is mortal

type.[6] His theory of syllogisms is subtle and complete, as far as it goes. But it does not amount to an adequate logic. Aristotle, in the writings which have survived, does not consider purely propositional reasoning, in which one handles whole propositions combined with 'and', 'or, 'not' and so on, rather the relationships between propositions which arise when they share names like 'Socrates' and predicates like 'mortal'.[7]

The Stoic philosophers who succeeded Aristotle developed propositional reasoning to a high level, almost up to the standard of George Boole's work in the nineteenth century. But their work is largely lost. From the time of the Stoics to the time of Boole little was achieved which could be of interest to us. This is apparent in the remarks of the late eighteenth-century philosopher Immanuel Kant who held that logic, since the time of Aristotle

has not been able to make one step in advance, so that, to all appearance, it may be considered as complete and perfect.[8]

For us, logic begins with the work of George Boole, a British mathematician, familiar to all computing scientists as the eponymous Boole of the booleans. Boole's work, set out in his *Laws of Thought* of 1854, resulted in that proto-logic which deals with purely propositional reasoning and which is what we call the propositional calculus. The propositional calculus handles logic at the level of a simple algebraic system, boolean algebra, the same algebra which we use to describe logic gates and to simplify logic circuits. As an analysis of reasoning it lacks power. Nevertheless, it shares with the predicate calculus (or more

---

[6] For a description of the nineteen valid forms of syllogism see Rudy Rucker's popular book *Mind Tools — The Mathematics of Information*, pp. 200-207.

[7] This omission in Aristotle's writings is a little odd since he himself provides an excellent example of purely propositional reasoning.

Either we ought to philosophise or we ought not.
If we ought, then we ought.
If we ought not, then we also ought.
Hence in any case we ought to philosophise.

See De Long *A Profile of Mathematical Logic*, 1970, p. 34ff for a discussion of Aristotle (and Kant).

[8] Again see de Long, *A Profile of Mathematical Logic*, 1970, p. 34ff.

strictly *first-order logic*) the fundamental features of having a syntax, a semantics, methods of proof theory and, lastly, a meta-theory which tells us about the relationship between these features. And so we study it as our first example of a system of formal logic. Logic proper, the predicate calculus, first-order logic, twentieth-century logic, was invented by the German mathematician Gottlob Frege, in 1879. It is the basic theory of mathematical logic. It incorporates the propositional reasoning of the Stoic philosophers, and also the syllogisms of Aristotle. It is also much stronger than both, providing as it does a system adequate for the logical development of almost all of mathematics. As the contemporary American philosopher and mathematician Willard van Orman Quine has written, clearly with Frege's achievement in mind

Logic is an old subject, and since 1879 it has been a great one.[9]

In the twentieth century there has been an explosion of interest in formal logic. Later in this book we develop formal systems of propositional and predicate logic as they have come to viewed by contemporary logicians. One of the greatest intellectual achievements of the twentieth century has been to exhibit both the power and the limitations of mathematical logic and of mathematics itself.

Mathematical logic has revealed both the power and the limits of logic and computability. Computers cannot do everything in principle, but it might seem at first sight that some of the things they cannot do, they ought to able to do. Logic teaches the computer scientist humility. For in the twentieth century the notion of *algorithm* has become the logicians' concern. Mathematicians deal in proofs. Logic provides a purely formal account of proof. If something is a proof, then a mechanical, uninventive intelligence can follow it. There is a mechanical method, or algorithm for checking a proof. (We all know how much easier it is to follow a proof than it is to invent one.) It is natural to ask whether or not there can be an algorithm for *generating* a proof if there is one to be found, and for reporting that there is no proof if there is none to be found, in some or all branches of mathematics. We should ask whether our belief that proof-following is so much easier than proof-making is in fact false. The question worried many famous logicians of the early part of the twentieth century. To answer it we need a clear conception of what an algorithm is.[10] In the

9 Quine, *Methods of Logic*, 1949, p. vii.

10 For a simple discussion of the notion of an algorithm see Goldschlager and Lister

course of finding one, and of answering the original question, the foundations for computing science were laid, by the logicians Alan Turing, Alonzo Church, Stephen Kleene and others. And the answer? There can be no algorithm, no computer program, for deciding pretty well any of the interesting questions of mathematics, an outcome which relieved mathematicians who would otherwise not only have been put out of business, but who would also have the joy taken out of their subject.

Things are even worse. Even when there is an algorithm, it may take more than the age of the Universe to execute on any existing machine. It may not be feasible. Finding our whether there is algorithm for the solution of some problem, and whether an algorithm is feasible if it exists, is one of the central topics of the theory of computation. These limitations on the computable are reflected in every programming language. They are especially evident in a language like Prolog whose formalism is itself modelled on that of formal logic.

1.4 Levels of Representation

Computing science is a long chain of connected topics, or sub-disciplines. Towards the bottom of the chain are descriptions of the events that occur in pieces of hardware which we represent as changes in bit-patterns stored in memory or in the registers of a processor. At the top of the chain come our representations of the computable in terms of logic and mathematics. The world at the bottom of this chain is a frenzy of activity, of things that change in time. At the top of the chain there is no change, only truth and falsity.

We connect mathematical logic, which describes computation at the highest, most abstract level, with what goes on at the lowest levels of representation with which the computing scientist is usually concerned, the level of bit-strings in a conventional computer. Computing science deals with very abstract objects like high-level programming languages, specification languages, systems of formal logic. It reaches down to the concrete physical objects like the hardware on which these higher level objects are implemented. If computation is to be intelligible to us humans, it must be appropriately chunked, and viewed hierarchically. We must conceptualize — that is, find appropriate descriptions of — what computing science deals with.

Programming on a 'higher' level has many advantages in addition to

Computer Science A Modern Introduction, 1982 p. 1ff.

portability. One of these is that we can make logical distinctions between types of data, between integers, characters and booleans, each of which has its appropriate and inappropriate range of associated operations. Another is less a matter of logic, more a matter of programming economy. The length of the source code for a given algorithm coded in a given language is inversely proportional to the level of the language. It is often said that studies show that programmers can produce correct, validated code at a certain rate, independent of the language in which they write. So the higher the level of language the greater the programmer's productivity. The move to higher levels of language is a means of coping with the complexity of software. In commercial terms it offers the prospect of a solution to the software crisis.

Pascal, designed by Niklaus Wirth in the early 1970s, is a distillation of all that it best in the Algol tradition. It is designed to promote the step-wise refinement of programs and structured programming. A year or two after its having been designed, it was re-defined axiomatically by Wirth and Hoare. Hoare had already developed the first logic of correctness for a portion of a real programming language. Associated with the axiomatic definition of Pascal by Wirth and Hoare is a logic of program correctness, further refined by Dijkstra and by David Gries among others. A logic of correctness can be made to support a method for developing software which proposes that programs be proved correct as they are developed. In the fully realized methodology of structured programming, logic becomes a practical tool for the programmer to use.

Here is a fragment of Pascal code which sums the natural numbers up to the natural number — that is, non-negative integer — **number**.

```
begin
{ number ≥  0 }
  sum := 0;
  count := 0;
  while count <> number do
  begin
        count := count + 1;
        sum := sum + count
  end
end; {of Pascal code}
```

When the loop terminates 'sum' is equal to the sum of the integers up to 'number'. Perhaps it is easy to see that summing the first N integers is what the program does. But how would you go about proving that that is what it does? How does a logic of program correctness work? In the method we develop in chapter 14, one marks with assertions what is true at each key point in the text of the program as it is executed. One treats the program text as a mathematical text about which one can state truths and reason logically. As a simple example, consider our 'sum_to_N' program. We know that initially 'number' is an integer greater than or equal to zero. And then we know that 'sum' and 'count' are both equal to zero after the first two assignment statements in the code. Immediately before the loop, dispensing with the quotes, we have

```
sum:=0;
count:=0;
{ number = N, number ≥ 0, sum = 0, count = 0 }
```

The documentation between the '{' and '}' is a list of assertions which are true when the processor executes the code in the textual position of the documentation. The loop is more difficult to handle. But one can show, and we shall in chapter 14, that the statement sum = $\sum_{j=0}^{j=count} j$ is always true both before and after each call of the body, the two assignment statements between the **begin** and **end,** of the loop. This statement is called the *invariant* of the loop, because it is always true both before and after each call of the loop body. When the loop terminates we know that, in addition, count equals number. So we can document the code as below.

{ number = N ≥ 0 }

sum:=0;

count:=0;{ number = N ≥ 0 , count = 0 , sum = $\sum_{j=0}^{j=count} j$ }

while count <> number do

begin {sum = $\sum_{j=0}^{j=count} j$ }

count:=count+1;

sum:=sum+count;

end;

{ number = N ≥ 0 , count = number , sum = $\sum_{j=0}^{j=count} j$ }

From the last of these assertions it follows that 'sum' contains the sum of the first N integers. To show that what you think is the loop invariant really is a loop invariant you need to know something about the logic of program correctness.To make the inference from the last assertion to the result you want requires a little logic.

'Sum_to_N' is a very simple program. Yet proving the correctness of even this program demands some work. One might ask why. The answer is that the program consists of a sequence of commands to the computer, commands which affect the contents of the computer's memory. Pascal, like most programming languages, is an imperative language. A program written in Pascal is essentially a sequence of commands. The relation of the commands in our program to what we want — the sum of the first N integers — is very indirect. The sequence of commands in our fragment of Pascal is very similar to the sequence of the statements as executed by a typical computer processor. The Pascal machinery used in our program is therefore low-level. But Pascal incorporates higher-level constructs. For example, it allows one to declare functions which may be called many times.

It allows these functions to be *recursive* - it allows the code in the body of the function to *call itself*. So 'sum_to_N' can be declared as the recursive function

```
function sum_to_N(number : integer) : integer;
  begin { function body }

        if number = 0
        then sum_to_N := 0
        else
        sum_to_N := number + sum_to_N(number - 1)

                        {function calls itself}

  end;{of function body }
```

which is an almost literal transcription of the specification which you mean to implement. Programming in a high-level language involves making fewer concessions to the underlying machine than to the human programmer. One's code can be relatively close to a formal specification of the algorithm one wants to implement.

1.5 Programming in Logic

What is the chance of programming our specification *directly* in to a programming language? In Prolog, a *non-imperative* language, a program consists of a set of logical statements. Each statement is either a fact or is a rule. Facts are categorical, rules are conditional. We can write a Prolog program to calculate sum_to_N. This consists of two statements, one a fact, the other a rule. Neither is a command. Each statement expresses a logical property of the predicate 'sum_to_N(N,S)' which, intuitively, is meant to state that the sum of the natural numbers up to and including N is S.

Here are our first Prolog programs. Read ':-' as *...if...*, and the commas ',' as *and*. Since we present Prolog as a purely logic programming language we shall have to build our own arithmetic from purely logical notions. So represent 0 by the numeral '**0**', and represent the successor function by s(·). Our first Prolog program contains four clauses. The first two deal with sum_to_N. The first of these says that the sum_to_N of the first 0 natural numbers is 0. The second says that the sum_to_N of the first **s(N)** natural numbers is **S**, if the sum_to_N of the first **N** natural numbers is **S1** and the result of adding **s(N)** to **S1** is **S**. The third and fourth clauses deal with the add predicate. They say that adding 0 to N gives N, and that adding the successor of N1 to N2 gives the successor of N3 if adding N1 to N2 gives N3.

```
sum_to_N(0,0).

sum_to_N(s(N),S) :-
  sum_to_N(N,S1),
  add(s(N),S1,S).


add(0,N,N).

add(s(N),N1,s(N2)) :-
  add(N,N1,N2).
```

This is 'arithmetic' of a very austere kind. We don't have the natural numbers we recognize, only 0 and its successors. Given the query[11]

```
| ? - sum_to_N(s(s(s(s(0)))),X).
```

the computer prints

```
X = s(s(s(s(s(s(s(s(s(s(0))))))))))
```

which says in effect that the sum_to_N of 4 is 10. Prolog is a programming language whose syntax is somewhat like that of predicate logic. It has an *inference engine* built into it. It is a higher-level language than Pascal because it conceals from us more of the operations of the computer. The Prolog code is much shorter than the Pascal code, which is itself of course much shorter than the machine code of the program. We can program more efficiently. We have fewer opportunities for making mistakes. Programming in Prolog is quite unlike programming in imperative languages and should ideally be preceded by a study of logic proper. But we live in the real world and make the following compromise. We introduce logic and Prolog in parallel, and then use Prolog to implement aspects of logic.

Because it is a simple very-high-level language and because it approximates to programming in logic, Prolog is coming to be the chosen language of the knowledge engineer, and the builder of expert systems, and some hope, that of the software engineer. But Prolog has intrinsic limits as a logic programming language. Because it is still relatively new, our examination of its failings will take us up to the state of the art in current research.

In the following chapters we will be concerned with the following questions:

What is logic and how can we formalize it?
To what extent can logic be automated?
What can logic do for the software engineer?
What does logic tell us about what computer programs mean?
What is logic programming and what are its limitations?

Summary

Computing science deals with representations of the computable. These

[11] Queries are discussed more fully in Chapter 2.

form a hierarchy extending from machine representations, which may be cast in the form of bit-patterns, through representations in higher-level languages like Pascal, through to representations in very high-level languages which are close to logic itself. Logic programming using the medium of Prolog seems to offer the possibility that we directly implement our logical specifications in code. It has become apparent that computing science not only has much to learn from mathematical logic but also much to contribute to it. Computing science threatens to revitalize mathematical logic. We now have logics which deal not with reasoning expressed in natural languages like English, but which deal with algorithms expressed in programming languages.

2 Prolog, pure and impure

Logic programming, as described in the previous chapter, a style of programming which allows the programmer to concern himself only with the declarative meaning of a program, is an ideal towards which computing scientists strive.

Prolog falls far short of this ideal, as we shall see. One reason for its falling short is that it possesses built-in predicates which have side-effects, among which are those which handle input/output and those which deal with arithmetic calculation. Another reason is that Prolog, as standardly implemented, enables the programmer to interfere with the control that the Prolog interpreter would otherwise impose on the execution of a program. Such direction of the Prolog interpreter, though undesirable in theory, is unavoidable in practice if reasonable efficiency of execution, and in some cases even termination, is to be guaranteed, as any Prolog programmer soon discovers for herself. For both of these reasons real Prolog is impure.

We study Prolog as an approximate logic programming language, that is, as an approximation to an implementation of formal logic. But we also use Prolog in the study of logic. We use Prolog to implement those aspects of formal logic, like parsing, proof-checking, and theorem-proving, which are most easily automated.

We need to some of the details of at least one standard Prolog implementation. The most standard is C-Prolog, also known as *Edinburgh* Prolog. The code we reproduce in our examples has standard C-Prolog or Edinburgh syntax, and should run on any standard implementation of the language.

We need to know something about the concrete syntax of Prolog — that is, something about the details of the texts which are to count as Prolog programs — and something about how the Prolog interpreter interprets a Prolog program. So we begin this chapter with a brief account of Edinburgh Prolog as a programming language, starting with a subset of Edinburgh Prolog which implements pure Prolog.

2.1 Pure Prolog

One aspect of Prolog is extremely simple. Prolog possesses just a single data type — the *term*.

Terms are either constants, or variables or are compound.

A constant is either a signed numeral (which stands for a number), or is an atom, in which case it is represented by a string of characters — which include the alphabetic characters, the digits, and various other characters like '?' etc. — which must begin with a *lower-case* letter.

A variable is represented by a string of characters beginning with an *upper-case* letter or with the underscore character.

A compound term consists of a functor (sometimes called a *predicate),* which is a constant, (say 'f') and its arguments, which are themselves terms. The *arity* of a functor is simply the number of arguments that it takes. If f has arity n, and t_1,...,t_n are terms, then f(t_1, ..., t_n) is a term.

Examples

Constants

0 -12 31876
jack
good_bye123

Variables

Hello
Good_bye
X
_X

Compound terms

hello(1,2,3)
good_boy(jack)

Every functor in Prolog has an arity, or number of argument places. The arity of **sum_to_N** in the previous chapter is 2, and of **add** is 3. In writing about the specific terms **sum_to_N** and **add** that we used in our program, we sometimes specify its arity along with its name. The convention is that we write **sum_to_N**/2 and **add**/3. A constant, say **0**, can be viewed as a degenerate compound term whose arity is zero. It is, in other words, a zero-ary functor.

There is a special functor which means '...if... '. It is written ':-'. So
':-'(good_boy(jack),will_have_cake(jack))
has ':-' as its functor, and says *if jack is good boy, jack will have cake.*

A Prolog program is a sequence of terms.

Each term in a Prolog program is an assertion, either a fact or a rule. So we sometimes refer to a Prolog program as a *database*. Facts, like **good_boy(jack)**, are simple and unconditional. But rules, like
':-'(good_boy(jack),will_have_cake(jack))
are compound and conditional rules are built out of fact-like components like **good_boy(jack)** and **will_have_cake(jack)** — terms which are either atoms or are compound terms whose functors are atoms. Rules like**':-'(good_boy(jack),will_have_cake(jack))** are most usually written

```
will_have_cake(jack) :-
        good_boy(jack).
```

The functor ':-' is treated as an operator, and written between its arguments, in *infix* style.[1]

Consider a simple short Prolog program — **sum.0** — for evaluating sum_to_N. The program contains five facts, and no rules.

```
                    /* sum.0 */

sum_to_N(0,0).
sum_to_N(s(0),s(0)).
sum_to_N(s(s(0)),s(s(s(0)))).
sum_to_N(s(s(s(0))),s(s(s(s(s(s(0))))))).
sum_to_N(s(s(s(s(0)))),s(s(s(s(s(s(s(s(s(s(0))))))))))).
```

How does one produce and use a Prolog program like **sum.0**?

First, one writes the source code — the collection of facts and rules — using an editor. On leaving the editor one calls the Prolog interpreter, and loads or *consults* the source code; that is, one loads the text so that the Prolog interpreter can execute it[2]. Finally, one

[1] For more on operators see chapter 5.

queries the consulted program. Text placed between the delimiters /* and */ is ignored by the Prolog interpreter when **sum.0** is loaded, or consulted.

Let us execute **sum.0** with a simple query. Here is a simple session, as it appeared on my terminal screen. What I, as user, typed is printed in *italic*. The computer output to the screen is printed in bold.

$ *cprolog*

C-Prolog version 1.5
| ?- *['sum.0'].*
sum.0 consulted 496 bytes 6.000012e-03 sec.

yes
| ?- *sum_to_N(s(s(s(0))),s(s(s(s(s(s(0))))))).*

yes
| ?- *trace,sum_to_N(s(s(s(0))),s(s(s(s(s(s(0))))))).*
(2) 1 Call: sum_to_N(s(s(s(0))),s(s(s(s(s(s(0))))))) ?
(2) 1 Exit: sum_to_N(s(s(s(0))),s(s(s(s(s(s(0)))))))

yes
| ?- *halt.*
[Prolog execution halted]

Commentary

First of all, I typed *cprolog* to enter the Prolog interpreter, and the screen displayed the message

C-Prolog version 1.5
| ?-

which told me I was running Prolog and should issue a query. So I consulted **sum.0** by typing *['sum.0'].* , and the interpreter replied with

sum.0 consulted 496 bytes 6.000012e-03 sec.

---

2 Which on my implementation is achieved by typing *cprolog* and consulting the program - the operating system is VMS. A shorthand for *consult(<file>)* is *[<file>]*.

yes
| ?-

which told me that **sum.0** had been consulted and therefore loaded as the program that was to be run. The screen told me how much space it took up and how long it took to load, and that the system expected a query. So I issued the query
sum_to_N(s(s(s(0))),s(s(s(s(s(s(0))))))).
which succeeded by matching with a fact in the program. The interpreter printed
yes
to the screen. Next I issued the query

trace,sum_to_N(s(s(s(0))),s(s(s(s(s(s(0))))))).

This query tells the interpreter to trace its execution — that is, to display its sequence of calls to the program and matches with the query. The comma between **trace** and
sum_to_N(s(s(s(0))),s(s(s(s(s(s(0)))))))
means 'and'. So

trace,sum_to_N(s(s(s(0))),s(s(s(s(s(s(0))))))).
means
trace the execution and execute the query
sum_to_N(s(s(s(0))),s(s(s(s(s(s(0))))))).

trace is the first example of a non-logical, built-in predicate, one which has a side-effect.

The trace
```
(2)    1 Call: sum_to_N(s(s(s(0))),s(s(s(s(s(s(0))))))) ?
(2)    1 Exit: sum_to_N(s(s(s(0))),s(s(s(s(s(s(0)))))))
```

yes
told me that **sum_to_N(s(s(s(0))),s(s(s(s(s(s(0)))))))** was called and succeeded immediately, causing an exit.
yes
| ?-
at which point I quit with
| ?-*halt.*
and the system replied with

[Prolog execution halted]
and I was out of Prolog.

Here is another session, this time in which a variable is used in the query.

$ *cprolog*

C-Prolog version 1.5
| ?- *['sum.0'].*
sum.0 consulted 496 bytes 4.999995e-03 sec.

yes
| ?- *trace,sum_to_N(s(s(s(0))),Sum).*
(2) 1 Call: sum_to_N(s(s(s(0))),_9) ?
(2) 1 Exit: sum_to_N(s(s(s(0))),s(s(s(s(s(s(0)))))))

Sum = s(s(s(s(s(s(0)))))) ;

(2) 1 Back to: sum_to_N(s(s(s(0))),_9) ?
(2) 1 Fail: sum_to_N(s(s(s(0))),_9)

no
| ?- *halt.*
[Prolog execution halted]

Commentary

This time I issued the query
trace,sum_to_N(s(s(s(0))),Sum).
using the variable 'Sum'. 'Sum' is translated[3] into the variable _9, so that the query becomes
sum_to_N(s(s(s(0))),_9)
which matches with the fourth fact in the database, so my original query succeeds with
Sum = s(s(s(s(s(s(0)))))) .

(2) 1 Call: sum_to_N(s(s(s(0))),_9) ?
(2) 1 Exit: sum_to_N(s(s(s(0))),s(s(s(s(s(s(0)))))))

[3] To avoid any possible variable clashes.

at which point I typed the semicolon *;* for further search through the database. But no other fact matched with the query so it failed and the system displayed

```
(2)      1 Back to: sum_to_N(s(s(s(0))),_9) ?
(2)      1 Fail:  sum_to_N(s(s(s(0))),_9)
```

no

| ?-

at which point I quit with

halt.

and the system replied with

[Prolog execution halted]

and I was out of Prolog again.

Exercise 2.1

Using **0**, and **s/1**, write Prolog clauses for

(a) **times/3**.
(b) **greater_than/2**
(c) **quotient_and_remainder/4**.
(d) **greatest_common_divisor/3**.

Neither of the previous examples exploits the use of variables in the database, though the second does have a variable in the query. So we return to the program of chapter 1.

/* sum.1 */

```
sum_to_N(0,0).

sum_to_N(s(N),S) :-
    sum_to_N(N,S1),
    add(s(N),S1,S).


add(0,N,N).

add(s(N),N1,s(N2)) :-
    add(N,N1,N2).
```

In the program **sum.1**
sum_to_N(0,0).
is a fact, and
```
sum_to_N(s(N),S) :-
    sum_to_N(N,S1),
    add(s(N),S1,S).
```
is a rule.

We read the rule as: *the sum_to_N of s(N) is S* if *the sum_to_N of N is S1* and *the add of s(N) and S1 is S*.

A rule consists of two parts, its *head* and its *body*. The body of the rule may be a single assertion, or it may be a conjunction of assertions. For example, in the rule

```
sum_to_N(s(N),S) :-
    sum_to_N(N,S1),
    add(s(N),S1,S).
```

sum_to_N(s(N),S)
is the head, and
```
    sum_to_N(N,S1),
    add(s(N),S1,S)
```

is the body.

The second rule for **sum_to_N**/2 exhibits a very important feature of Prolog, that rules are frequently *recursive*. The predicate in the head often re-appears in the body (though with different arguments). When the Prolog interpreter calls **sum_to_N(N,S1)** it then has to call **sum_to_N(N,S1)**, and so on down to **sum_to_N(0,0)**. Each of these calls is recursive, and successive calls are of greater recursive *depth*. Given this program, here is a trace of the query

| ?- *sum_to_N(s(s(s(0))),Sum).*

The un-bracketed numbers which come after the bracketed numbers at the beginning of each line of the trace represent the depth of the recursive call. I have edited the trace so that deeper calls are shifted to the right.[4]

| ?- *trace,sum_to_N(s(s(s(0))),Sum).*

```
(2) 1 Call:   sum_to_N(s(s(s(0))),_9) ?
(3) 2 Call:       sum_to_N(s(s(0)),_65612) ?
(4) 3 Call:           sum_to_N(s(0),_65622) ?
(5) 4 Call:               sum_to_N(0,_65632) ?
(5) 4 Exit:               sum_to_N(0,0)
(6) 4 Call:               add(s(0),0,_65622) ?
(7) 5 Call:                   add(0,0,_25) ?
(7) 5 Exit:                   add(0,0,0)
(6) 4 Exit:               add(s(0),0,s(0))
(4) 3 Exit:           sum_to_N(s(0),s(0))
(8) 3 Call:           add(s(s(0)),s(0),_65612) ?
(9) 4 Call:               add(s(0),s(0),_29) ?
(10) 5 Call:                      add(0,s(0),_33) ?
(10) 5 Exit:                      add(0,s(0),s(0))
(9) 4 Exit:               add(s(0),s(0),s(s(0)))
(8) 3 Exit:           add(s(s(0)),s(0),s(s(s(0))))
(3) 2 Exit:       sum_to_N(s(s(0)),s(s(s(0))))
(11) 2 Call:          add(s(s(s(0))),s(s(s(0))),_9) ?
(12) 3 Call:              add(s(s(0)),s(s(s(0))),_37) ?
(13) 4 Call:                  add(s(0),s(s(s(0))),_41) ?
(14) 5 Call:                      add(0,s(s(s(0))),_45) ?
(14) 5 Exit:                  add(0,s(s(s(0))),s(s(s(0))))
```

[4] On the screen, the trace appears without the shifts to right and left.

```
(13) 4 Exit:
   add(s(0),s(s(s(0))),s(s(s(s(0)))))
(12) 3 Exit:
   add(s(s(0)),s(s(s(0))),s(s(s(s(s(0))))))
(11) 2 Exit:
   add(s(s(s(0))),s(s(s(0))),s(s(s(s(s(s(0)))))))
(2) 1 Exit:   sum_to_N(s(s(s(0))),s(s(s(s(s(s(0)))))))

Sum = s(s(s(s(s(s(0))))))
yes
```

Commentary

After the conversion of the variable 'Sum' the literal **sum_to_N(s(s(s(0))),_9)** matches with, or is instantiated to, the first literal in the rule for sum_to_N. Similarly **sum_to_N(s(s(s(0))),_9)** matches with **sum_to_N(s(s(0)),_9)** and with **sum_to_N(s(0),_9)** and so on down to **sum_to_N(0,_9)** . Thus _9 is instantiated to **0** at the fourth call and exits. At this point the fact and rule for **add**/3 are called 4 times exiting with **sum_to_N(s(0),s(0))** at the tenth line of the trace.

The process continues with 6 calls to **add**/3 exiting with **sum_to_N(s(s(0)),s(s(s(0))))** at line 17, and then with 8 calls to **add**/3 which finally yields
sum_to_N(s(s(s(0))),s(s(s(s(s(s(0)))))))
at line 26.

Re-ordering the terms in the body of the rule for **sum_to_N**/2 we obtain

```
sum_to_N(0,0).

sum_to_N(s(N),S) :-
   add(s(N),S1,S),
   sum_to_N(N,S1).          /* inverted order of literals */
```

and the query traces as follows. Notice that the order of the calls to **add**/3 is reversed, as are the calls to **sum**/3.

| ?- *trace,sum_to_N(s(s(s(0))),Sum).*

```
(2) 1 Call:  sum_to_N(s(s(s(0))),_9) ?
(3) 2 Call:      add(s(s(s(0))),_65612,_9) ?
(4) 3 Call:          add(s(s(0)),_65612,_23) ?
(5) 4 Call:              add(s(0),_65612,_27) ?
(6) 5 Call:                  add(0,_65612,_31) ?
(6) 5 Exit:                  add(0,_31,_31)
(5) 4 Exit:              add(s(0),_31,s(_31))
(4) 3 Exit:          add(s(s(0)),_31,s(s(_31)))
(3) 2 Exit:      add(s(s(s(0))),_31,s(s(s(_31))))
(7) 2 Call:      sum_to_N(s(s(0)),_31) ?
(8) 3 Call:          add(s(s(0)),_65657,_31) ?
(9) 4 Call:              add(s(0),_65657,_36) ?
(10) 5 Call:                     add(0,_65657,_40) ?
(10) 5 Exit:                     add(0,_40,_40)
(9) 4 Exit:              add(s(0),_40,s(_40))
(8) 3 Exit:          add(s(s(0)),_40,s(s(_40)))
(11) 3 Call:             sum_to_N(s(0),_40) ?
(12) 4 Call:                 add(s(0),_65693,_40) ?
(13) 5 Call:                     add(0,_65693,_45) ?
(13) 5 Exit:                     add(0,_45,_45)
(12) 4 Exit:                 add(s(0),_45,s(_45))
(14) 4 Call:                 sum_to_N(0,_45) ?
(14) 4 Exit:                 sum_to_N(0,0)
(11) 3 Exit:             sum_to_N(s(0),s(0))
(7) 2 Exit:      sum_to_N(s(s(0)),s(s(s(0))))
(2) 1 Exit:  sum_to_N(s(s(s(0))),s(s(s(s(s(s(0)))))))

Sum = s(s(s(s(s(s(0))))))
yes
```

Logical connectives in pure Prolog

Prolog, though at least a first approximation to a logic programming language, suffers from a paucity of logical connectives, or logical operators. It has a conjunction (an *...and...*), represented by the comma, as we have seen. It also has a conditional (an *...if...*) represented by the symbol ':-'. Prolog imposes the important restriction that the head of a rule — that is its consequent, the part before the 'if' — must be an atom or a compound term whose functor is an atom. Prolog does have a disjunction (an *...or...*) of a restricted sort, represented by ';'. One's use of the Prolog disjunction in a Prolog program or database is restricted to the *body* of a rule. So that

```
p :- (q ; r).
```

is equivalent to the two rules

```
p :- q.
p :- r.
```

In a Prolog program, one cannot have a disjunctive 'fact' ('either p or q ')

```
p ; q.                          /* error !!! */
```

and one cannot have a rule whose head is disjunctive ('p or q, if r ')

```
(p ; q) :- r.                   /* error !!! */
```

Try putting either of these into a Prolog program and Prolog will give you an error message when you try to load (or 'consult') it.[5]

Queries are terms which are presented to a Prolog program. One of three things can happen to them. They either *succeed*, or *fail*, or do neither in which case the program does not terminate. This last is an outcome one usually wants to avoid. Ideally, a query succeeds if and only if it is logically implied by the program, which is, as we said, a collection of assertions. So how is the execution mechanism revealed by **trace/0** really handled in Prolog?

2.2 The Prolog inference engine: backtracking

Here is a simple Prolog program which has two rules and six facts.

```
                    /* backtracking example */

/* rules */

    p(X) :-
            q(X),
            r(X).

    q(Z) :-
```

[5] This restriction is one of the ways in which Prolog falls short of being an implementation of pure logic. It is the restriction imposed by Prolog's implementing not the full classical logic that we study later in this book, but the much weaker *Horn Clause Logic*, whose properties we discuss in chapter 15.

```
        s(Z).

/* facts */

    q(a).
    q(b).

    r(b).
    r(c).
    r(d).

    s(d).
```

The first rule in **backtracking example** is

```
p(X) :-
    q(X),
    r(X).
```

This asserts that *for all X*, p(X) is the case, if both q(X) and r(X) are the case. A variable X is interpreted throughout a fact or a rule in which it occurs in a program as meaning *for all X*. **a**, **d**,**c**, and **d** are constants. The six facts read *q(a) is case ,... s(d) is the case*.

If we want to discover which constants the predicate **p**/1 is true of we consult the program, and issue the query | **?- p(X).** But before doing that we can consider which answers Prolog ought to deliver.

For example, should Prolog confirm that **p(a)** is the case? The answer is that it should if **q(a)** and **r(a)** are both the case. **q(a)** is the case, because it is given as a fact. But **r(a)** is not the case. It is not given as a fact, nor does it follow from either of the rules. So Prolog ought not to show that **p(a)** is the case.

What about **p(b)**? **q(b)** is the case — it is the second fact in the program, so **p(a)** will be the case if **r(b)** is the case, according to the first rule in the program. **r(b)** is also given as a fact. So Prolog should show that **p(b)** is the case.

What about **p(c)**? **q(c)** is not given as fact, and since **s(c)** is not given as a fact, **q(c)** does not follow from the second rule. So Prolog ought not to show that **p(c)** is the case.

Finally, what about **p(d)**? **q(d)** is not given as a fact, but it does follow from **s(d)** and the second rule. **r(d)** is given as a fact, and so Prolog should show that **p(d)** is the case.

How does Prolog actually behave?
Here is a trace of the query | ?- p(Y).

```
| ?- trace,p(Y).
 (2) 1 Call:   p(_0) ?
 (3) 2 Call:          q(_0) ?
 (4) 3 Call:                 s(_0) ?
 (4) 3 Exit:                 s(d)
 (3) 2 Exit:          q(d)
 (5) 2 Call:          r(d) ?
 (5) 2 Exit:          r(d)
 (2) 1 Exit:   p(d)

Y = d ;

 (5) 2 Back to:              r(d) ?
 (5) 2 Fail:          r(d)
 (4) 3 Back to:                     s(_0) ?
 (4) 3 Fail:                 s(_0)
 (3) 2 Exit:          q(a)
 (6) 2 Call:          r(a) ?
 (6) 2 Fail:          r(a)
 (3) 2 Back to:               q(_0) ?
 (3) 2 Exit:          q(b)
 (7) 2 Call:          r(b) ?
 (7) 2 Exit:          r(b)
 (2) 1 Exit:   p(b)

Y = b ;

 (7) 2 Back to:              r(b) ?
 (7) 2 Fail:          r(b)
 (3) 2 Back to:              q(_0) ?
 (3) 2 Fail:          q(_0)
 (2) 1 Back to:       p(_0) ?
 (2) 1 Fail:   p(_0)

no
```

Why does the Prolog interpreter deliver these solutions in this

particular order? To understand why it does one has to understand how the Prolog interpreter implements its *search mechanism* and how it uses *backtracking*. The query | ?- *trace,p(Y).*
succeeds first with

Y = d.

The line

(5) 2 Back to: r(d) ?

signals that backtracking takes place after the call for more solutions to the query. Then with

Y = b

after which it fails. That is, given the query Prolog returns

Y = d

after which the user types a semicolon to obtain the further solution(s). The only further solution is

Y = b

Prolog searches for a match to the query | **?- p(Y).** by examining the program from top to bottom and from left to right, in the following sense. First **p(Y)** is matched with the head of the first rule. **p(Y)** will succeed if the **Y** in **q(Y)** and **r(Y)** may both be *instantiated* to some term occurring in the program.

So Prolog then sets out to instantiate the **Y** in **q(Y)**, if it can. Its first successful match, again starting from the top of the program, is the second rule. So the **Y** in **q(Y)** will be instantiated to whatever **Y** in **s(Y)** is instatiated to, if anything. Prolog now searches the program, again from the top, to see to which terms the **Y** in **s(Y)** may be instantiated.

Having instantiated the **Y** in **q(Y)** with **d**, Prolog then moves rightwards along the first rule to see if **r(d)** matches a term in the program, and it does this by matching with fifth fact in the program. By this time Prolog has exhausted its possibilities for instantiating the **Y** in **q(Y)** via the second rule, and so it must work through the facts. At that point **p(Y)** succeeds by instantiating **Y = b**.

2.3 Impure Prolog: Built-In Predicates

Pure Prolog is a very austere programming formalism. Standard implementations of the language incorporate a range of built-in predicates whose behaviour is standardized (more or less) and whose use either makes Prolog more user-friendly or, in some cases, makes its use practicable where the use of Pure Prolog is not.

Perhaps the most pervasively used built-in predicate is equality, the identity =/2.

Equality

Given the query
| ?- **=(X,0).**
the Prolog interpreter succeeds with
X = 0

Similarly, given the query
| ?- **=(0,X).**
the Prolog interpreter succeeds with
X = 0

Actually = is both a predicate and an operator. Operators may be written in *infix* style, and = usually is. So the query
| ?- **X = 0, X = Y.**
the Prolog interpreter succeeds with
X = 0
Y = 0

Pure Prolog manages with pattern-matching and nothing else. However, even if we want to avoid *computing by effect,* we still require side-effects to write to the screen, to read from the terminal keyboard, to write to and read from files in general.

Input/output

Prolog provides a very low-level procedure which writes a *single character* to the output, whether this is a screen or a file. The procedure is **put**/1. It always succeeds. Given the integer argument X, the character whose ASCII code is X is written to the output. The functor **put**/1 may appear in a query, or in the body of a clause held in a program. The query

| ?- **put(97).**

succeeds and writes 'a' to the output

a

yes

The simplest procedure with a desirable side-effect, is the *new-line* procedure, or **nl**/0, which simply writes a line-feed to the output.

For example

```
| ?- put(97),nl.
```

succeeds and writes 'a' to the output

a

yes

The converse case, reading a single character from the input, is managed by two built-in predicates **get**/1 and **get0**/1. **get0(X)** reads the next character from the input and instantiates X with its ASCII code. **get(X)** does the same, except that it ignores non-printing characters. Files are handled by changing the current input and output streams. Thus if, instead of reading a character from the keyboard, we want to read it from a file called 'file_name' we can use the query

```
| ?- seeing(Old),see('file_name'),get0(X),seen,see(Old).
```

If the file named *file_name* begins with the word 'hello', then X is instantiated to the ASCII code for 'h'. So the query succeeds with

```
Old = user
X = 104;
yes
```

seeing(Old) is instantiates **Old** to the original input stream. This initially comes from the terminal.

see('file_name') opens the file named 'file_name'. get0(X) gets the ASCII code of the first character. **seen** closes 'file_name', and **see(Old)** returns the current input stream to the terminal. There are corresponding built-in predicates for output. These are **telling**/1,**tell**/1,and **told**/0. Here is a Prolog program which reads the characters in a file and writes then to the current output stream.

Finally, most Prolog implementations have a built-in constant **end_of_file**/0 which matches with the end of file code, as used in the following example.

```
                /* file example 1 */

print_file(File) :-

        seeing(X),
        see(File),
        read_and_put_char,
        seen,
        seeing(X).


read_and_put_char :-

    get0(X),
    (    X = end_of_file
         ;
         put(X),
         read_and_put_char ).
```

This program has the disadvantage that every call to **get0(X)** involves a recursion via **read_and_put_char**. One can display only short files in this way. Soon execution will fail owing to a stack overflow.

An alternative is to replace the recursion embodied in **read_and_put_char** with an explicit iteration, using a built-in predicate **repeat** together with a device called **cut**. The cut is the subject of the next section. **put**/1, **get**/1, and **get0**/1 deal with I/O at the level of the single. Prolog lets one deal with I/O at the level of whole terms, using **read**/1 and **write**/1. For example, the query

```
| ?- read(X).
```

causes input to be read, and the variable X is then instantiated to it

```
| 1 + 2.
X = 1 + 2
```

Write/1 writes terms to the output stream.

Exercise 2.2

Write a Prolog program which reads text from a file and which writes the file directly to the output stream changing lower-case letters to upper-case letters.

Changing the database

Prolog allows the user to alter the Prolog database — or program — while it is running. The built-in predicate **asserta**/1 takes a Prolog clause — usually a fact — as its argument and adds it to the start of the database. The built-in predicate **assertz**/1 takes a Prolog clause — usually a fact — as its argument and adds it to the *end* of the database. **retract**/1removes the first occurrence of a clause from the database.

asserta/1, **assertz**/1, and **retract**/1 are to be used with care. They clearly spoil the logical clarity of a Prolog program, if it is viewed as a *logic* program. But we shall use them first in chapter 6.

Arithmetic

C-Prolog does allow the more familiar use of the numerals as in our second version of the program for sum_to_N. This uses the **is** of evaluation and the '+' operator which represents addition. But the price we pay by adding these arithmetical primitives to Prolog is that Prolog ceases to be a purely logic programming language.

Here is a version of sum_to_N which uses the arithmetic resources of Prolog as it is actually implemented.

```
sum_to_N(0,0).

sum_to_N(N,S) :-
   sum_to_N(N1,S1),
   N is N1 + 1,
   S is S1 + 1.
```

The first statement tells us that the sum of the first zero natural numbers is zero. The second statement tells us that the sum of the first **N** natural numbers is **S** if the sum of the first **N1** natural numbers is **S1**, where **N** is **N1** plus **1**, and **S** is **S1** plus **N**.

This is just what we mean by the sum of the natural numbers up to and including N. And so, to a good approximation, Prolog allows us to write our specifications directly into code Notice how much shorter the Prolog program is than the equivalent Pascal program. To the tutored eye, the Prolog program also has a more perspicuous logic.

If, running Prolog, we query

```
| ? - sum_to_N(10,X).
```

the computer will print

```
X = 55
```

2.4 Impure Prolog: the cut !

The cut — written **!** — is a predicate built-in all implementations of Prolog. The use of cut is one of the most important ways in which a Prolog program will fall short of being a true logic program. If you use cut, you damage purely logical, or the declarative, interpretation of your program. On the other hand, efficiency often demands that the cut be used, as we shall see.

The cut is a *zero-ary predicate* which always succeeds. Another predicate which always succeeds is the built-in predicate **true**. **false**, on the other hand, always fails. Since it is built-in, it cannot appear in the head of a Prolog clause. It can only appear in the body of a rule, or in a query. What does the cut do?

The effect of the cut is to succeed and to trim (or cut) the processes

(1) of calling further clauses in a procedure, and

(2) of backtracking.

Take the first case — that of calling further clauses in a procedure — first. Suppose you consult the next program **cut example 1**, and issue the query that follows:

/* cut example 1 */

```
p(a) :- !.

p(b).
```

|- ? *p(Z).*

The query will clearly succeed with **Z = a**. Without the cut, the query could then succeed with **Z = b**. But in this case, cut prevents further matching of the query with the second clause. So that after succeeding with **Z = a**, the query fails. So one has

```
|- ? p(Z).
Z = a;
no
```

One obtains identical behaviour if, instead of two separate clauses, a disjunction is used, as in **cut example 2**.

/* cut example 2 */

```
p(a) :-
   (!
    ;
    p(b)).
```

However, consider the case of **cut example 3**. Here a query of **p(Z)** succeeds first with **Z = a**, because **q(X)** succeeds with **q(a)**. The cut prevents a further call to **q(b)**. So the query **p(Z)** can then succeed only with **Z = c**.

/* cut example 3 */

```
p(X) :-
   q(X).

q(a) :- !.

q(b).

p(c).

   |- ? p(Z).
   Z = a;
   Z = c;
   no
```

In a *query*, cut works simply to restrict backtracking to further solutions. For example, consider the following program

/* cut example 4 */

```
p(a) .
p(b).

q(a).
q(b).
```

The query
| ?- *p(X),q(Y).*
succeeds with

```
X = a
Y = a;
X = a
Y = b;
X = b
Y = a;
X = b
Y = b;
no
```

All possible of combinations of instantiations for X and Y are delivered. However, the query

| ?- *p(X),!,q(Y).*

succeeds with only
X = a
Y = a*;*
X = a
Y = b*;*
no

In this case the cut prevents more than one instantiation for X. But Y is still free to be instantiated by both a and b. And the query
| ?- p(X),q(Y),!.
succeeds with only
X = a
Y = a*;*
no

In this case both X and Y are constrained to be instantiated only once, with a each time.

The cut-free query
| ?- *p(X),q(X).*
succeeds with
X = a*;*
X = b*;*
no

While the queries
| ?- *p(X),!,q(X).*
and
| ?- *p(X),q(X),!.*
both succeed with only
X = a*;*
no

Here is another use of cut, to prevent backtracking once the end of a file is reached.

```
                    /* file example 2 */

print_file(File) :-

        seeing(X),
        see(File),
        repeat,
        get0(Y),
        put(Y),
        Y = end_of_file,
        !,                              /* cut ! */
        seen,
        seeing(X).
```

2.5 Lists In Prolog

Lists form a very important data-type in high-level languages like Lisp and Prolog. A list is a sequence. It is unlike a set in that the order of elements of a list is significant.[6] The empty list is written as [], and is pronounced 'nil'. If we take a list List, and want to place a term Term at its head, we can write

.(Term,List)

The dot '.' is a built-in functor of Prolog. .(Term,List) is more conveniently written

[Term | List]

and lists are usually denoted by sequences of elements separated by commas written within square brackets. [1,4,3,2,5] is a list whose elements are natural numbers. It has five elements. The query

| ? - *[Head / Tail] = [1,4,3,2,5].*

succeeds with

Head = 1
Tail = [4,3,2,5]

The first element of a list is quite generally called its *head*, and the rest, its *tail*.

Lists are extremely useful in Prolog programming. We should mention two slightly odd built-in predicates of Prolog which are sometimes useful in manipulating lists. These are **=..** and **functor**/3. =..

[6] For more on all this see chapter 3.

is an operator.[7]

```
$ cprolog
C-Prolog version 1.5
| ?- [Head|Tail] = [[a,[b]],c,[d],e].
Head = [a,[b]]
Tail = [c,[d],e] ;
no

| ?- [Head|Tail] = [].
no

| ?- X =.. [disliked,ludwig,bertie].
X = disliked(ludwig,bertie) ;
no

| ?- functor(loved(bertie,wife('alfred north')),Functor,Arity).
Functor = loved
Arity = 2 ;
no

| ?- name(bertie,List).
List = [98,101,114,116,105,101] ;
no
```

Exercise 2.3

Write Prolog clauses for **occurs_in**/2, which succeeds if and only if its first argument occurs as an element of the list occurring in the second argument place.

Summary

In this chapter we examined the basic syntax of Pure Prolog — Prolog

[7] An operator is nothing more a functor which may be written between its arguments rather than in front of them. However, because of this syntactic peculiarity a (binary) operator must have a precedence and an associativity (for which see chapter 4).

without built-in predicates and the cut. We saw that Prolog possesses a single data type, the term. A Prolog program is a sequence of special terms — facts and rules.

We examined the mechanism of Prolog evaluating of queries, using pattern-matching and backtracking.

Finally we looked at how some the built-in predicates of Prolog may be used for input/output and arithmetic.

3 Elementary mathematics in Prolog

In this chapter we solve some more interesting programming problems using Prolog. We begin with some preliminary discrete mathematics. The mathematical ideas we need in order to handle formal logic are, unsurprisingly, just those ideas that we need in order to understand the business of computer programming. Discrete mathematics and the theory of formal grammars underpin the syntax and the semantics of both mathematical logic and programming languages. To handle the semantics of logic we need to understand of sets, functions, and relations. To handle the syntax of logic we need to understand what strings, lists, and trees are, as well as formal grammars and formal languages.

So we begin with the mathematical idea of sethood, move on through ordered pairs, n-tuples, functions, S-expressions, lists, strings, trees, and in chapter 4, formal grammars, and formal languages.

3.1 Sets

The ideas of sethood and set membership of a set have a good claim to be the most fundamental in all of mathematics. Sets are *the* fundamental mathematical objects in the precise sense that all mathematical objects — objects as varied as numbers, algebras, groups, lines, topologies, geometries, however we imagine them informally — can be construed, or re-constructed, as sets.

A set is a collection of objects, a new entity which is *distinct from the objects which are its members*. Take any fixed collection of objects a, b, c, d .... We can form sets of objects taken from this collection. Thus, if

$$A = \{a,b,d\}$$

then A is a set with the three members a, b and d. If we write that

B = {a,d,b,a,b}

then we again have a set with three members a, b, and d. Order and repetition simply make no sense when applied to sets. Sets are simple unstructured entities, simple unstructured collections. The ideas of order and do repetition apply to our *descriptions* of sets. Our description of a set is quite a different thing from the set we describe.[1] In our example, A and B happen to be the same set: what defines a set is its members. Put paradoxically: two sets with the same members are the same set. We write 'x is a member of A', and 'x belongs to A' as 'x ∈ A'.

So we may state the identity condition for sethood. For sets A and B

A = B	$\triangleq$	for all x,	both x ∈ A whenever x ∈ B,
			and x ∈ B whenever x ∈ A.

The symbol $\triangleq$ means *is defined to be the same as.* You will find the same symbol in many books on computing science. We also use the abbrevation '...iff...' to mean '...if and only if...', where what is intended is a statement of fact (or of mathematical fact) but is not taken to be a definition. Sets are *extensional objects.*[2]

The set A is the same set as set B if and only if first, all members of A are members of B, and secondly, all members of set B are members of set A. If every member of a set A is a member of set B, then we say that

[1] Never confuse the description with the thing described ! In defining the set A we use a symbol 'A' which names the set A, and we also use the symbols '{', 'a',',','b','d', and '}'. The opening '{' and closing '}' tell us that what occurs in between characterizes the set. The commas are merely delimiters - like semicolons in Pascal - and the 'a','b','d' are names of three distinct objects. Names are linguistic objects - words or symbols. They refer to objects, in fact they refer to the objects they name. Notice that we assume that none of the objects under discussion has more than one name, otherwise we could not be confident that the set A has exactly three members. All this might seem very laboured, especially when dealing with something as apparently simple as the idea of sethood. But it is important in computer science to be clear about the difference between *representation* and *object represented*, the difference between words and things. Sets are unordered. But any representation of set which lists its members will have an ordering.

[2] They are defined by their 'extent' rather than by the means we use to specify them. So, according to the classic example due to Bertrand Russell, the set of human beings is the same set as the set of featherless bipeds, though *human* and *featherless biped* are quite different properties. This is a simple idea, but the English sentence which expresses it looks cluttered and potentially ambiguous. One of the many bonuses of a formal treatment of logic is that it enables one to express both simple and complex facts tersely and without ambiguity, as we shall see when, in chapter 11, we have special symbols for *for all...*, *...whenever...* and *...and...* .

A is a *subset* of B and that B is a *super-set* of A.

We write this fact $A \subseteq B$. From the definition of set identity it follows that

$$A = B \text{ iff } A \subseteq B \text{ and } B \subseteq A.$$

Any set may be *named,* but most sets are too large to specified by a sequence of names of their members. Instead they must be specified by some property that all the members of the set possess. A simple example is the set $\mathbb{N}$ of natural numbers.[3]

We represent the set S of all objects which have some property ψ in common by

$$S = \{ x : \psi(x) \}$$

Of course, we could substitute any 'y' for 'x' and write, equally well,

$$S = \{ y : \psi(y) \}$$

So, the union $A \cup B$ of two sets A and B is the set which has as its members exactly those objects which belong to either A or to B (or to both A and B). Thus

$$A \cup B = \{ x : x \in A \text{ or } x \in B \}$$

The intersection $A \cap B$ of two sets A and B is the set which has as its members exactly those objects which belong to both A and to B.

$$A \cap B = \{ x : x \in A \text{ and } x \in B \}$$

When two sets A and B have no members in common, then their intersection is the empty set and they are said to be *disjoint*.

3.2 Ordered pairs

Sets are completely lacking in structure. Often we need entities which are like sets but which are ordered. Take two different natural numbers x and y. The set $\{x, y\}$ is the same set as $\{y, x\}$, and both are good representations of the 'unordered pair of x and y'. But sometimes we want the 'ordered pair of x and y', represented as $<x, y>$.

Ordered pairs are elements of relations and functions. We want to be able to construe relations and functions extensionally as sets. For example, the function **successor** embodies a rule which associates a natural number with another natural number, namely the next one. But

[3] We define $\mathbb{N} = \{0, 1, 2, 3,..... \}$. Incidentally, some writers omit 0 from the natural numbers assuming, one supposes, that 0 is less natural than 1 and 2 and ....
We follow the 'less natural' convention. But we denote the natural numbers other than zero by $\mathbb{N}^+ = \{1, 2, 3,..... \}$.

the idea of a function as a 'rule' associating pairs of values is obscure. We shall want to say that the rule associated with **successor** can be replaced by a set of ordered pairs each having the property that the second element of the pair is the square of the first. We want to be able to say that $<0, 1>$ 'belongs to' square, but that $<1, 0>$ doesn't. We shall want to say that

successor $= \{<x,y> : x \in \mathbb{N}, y \in \mathbb{N}, \text{and } y = x + 1\}$

so that

successor $= \{ <0,0>, <1,1>, <2,4>, <3,9>,....\}$

In fact there are many ways of defining the 'ordered pair' of two objects in terms of sets containing those two objects. What we must do is to *choose* a representation of $<x, y>$ in terms of x, y and sets of x and y (and perhaps sets of sets of x and y, and so on) so that from

$$<x_1, y_1> = <x_2, y_2>$$

we can infer that $x_1 = x_2$ and that $y_1 = y_2$, an inference we cannot make if we represent $<x, y>$ as $\{x, y\}$. The usual solution to the problem is to choose the representation

$$<x, y> = \{ \{x\}, \{x, y\} \}$$[4]

3.3 Binary relations and functions

A binary relation like *...is less than...* is something that holds between pairs of objects. We can think of ...less than... as a set of ordered pairs whose elements we conventionally restrict to $\mathbb{N}$, so that *...is less than...* is the set

$$\{ <0,1>, <0,2>,...., <1,2>, <1,3>,..., <2,3>, <2,4>,.... \}$$

...is less than... is a binary (or, less elegantly, 2-ary) relation. The set of elements which can figure as the first element of one of the ordered pairs in the relation is its *domain*, and the set of elements which can figure as the second element of one of the ordered pairs is its *co-*

[4] We can prove that, according to this representation, the desired result that $<x_1, y_1> = <x_2, y_2>$ implies that $x_1 = x_2$ and $y_1 = y_2$. There are two cases to consider. Either $x_2 = y_2$ or $x_2 \neq y_2$. Suppose, on the one hand, that $x_2 = y_2$. Then $<x_2, y_2> = \{\{x_2\}, \{x_2, y_2\}\} = \{\{x_2\}, \{x_2, x_2\}\} = \{\{x_2\}\}$. Therefore $\{\{x_1\}, \{x_1, y_1\}\} = \{\{x_2\}\}$. And so $x_1 = x_2$ and $y_1 = x_2 = y_2$. Suppose, on the other hand, that $x_2 \neq y_2$. It follows that $\{x_2\} \neq \{x_2, y_2\}$, since the first set has one member and the second has two. Similarly $\{x_1\} \neq \{x_2, y_2\}$. It follows that $\{x_1\} = \{x_2\}$, and further that $\{x_1, y_1\} = \{x_2, y_2\}$, so that $x_1 = x_2$ and $y_1 = y_2$.

domain. The domain and co-domain of a two-place relation need not be the same set, though they are in this example.

We define the *Cartesian product* A × B, of sets A and B, as the set of ordered pairs whose first element belongs to A, and whose second element belongs to B. Thus

$$A \times B = \{\langle a, b\rangle : a \in A \text{ and } b \in B\}$$

A binary, relation *between* A and B is then a subset of the Cartesian product A × B. A binary relation *on* A is a subset of the Cartesian product A × A.

Example

Ttake

$$A = \{1,2,3\}$$
$$B = \{1,2\}$$

then the Cartesian product

$$A \times B = \{ \langle 1,1\rangle, \langle 1,2\rangle, \langle 2,1\rangle, \langle 2,2\rangle, \langle 3,1\rangle, \langle 3,2\rangle \}$$

and the relation $\leq_{A \times B}$, *less than or equal to between A and B* is

$$R = \{ \langle 1,1\rangle, \langle 1,2\rangle, \langle 2,2\rangle \}$$

A function is a special binary relation which associates a member of a set — the member is its argument — with another object — its value for that argument — a member of a set, which may be the same set or a different set. A function has the property that for a given argument, there is only one value. Therefore *...is less than...* is not a function, since both <0,1> and <0,2> etc. belong to it.

A function is therefore associated with two sets, the first again called its *domain*, which is the set of all objects it applies to, the set of all its arguments, and the second again called its *co-domain*, a set which is a super-set of the set of all its values. The set of all its values, a subset of its co-domain, is called its *range*.[5] As we noted, the domain and the co-domain of a given function are sometimes the same set, and sometimes not. Sometimes every member of the co-domain is the value of the function for one of its arguments, sometimes not. Strictly speaking, a function is defined by the rule which associates argument with value, together with its domain and its co-domain. A function can have any set

[5] So that the domain and co-domain of the function **successor** is $\mathbb{N}$, but its range is $\mathbb{N}^+$.

as its domain. But change the domain of a function, or enlarge its co-domain and you have a different function. Thus we identify the function **successor**

$$\mathbf{successor}(n) = n + 1, \text{ for all } n \in \mathbb{N}$$

with three things:

(1) its domain, $\mathbb{N}$;

(2) its co-domain, $\mathbb{N}$ and

(3) the set $\{ <0,1>, <1,2>, <2,3>, \ldots\ldots\} \subseteq \mathbb{N} \times \mathbb{N}$

Often, when the domain and the co-domain of the function can be taken as understood, we speak as if a function were simply (3), namely the set of ordered pairs of arguments and values. Of course, functions can have domains and ranges other than the natural numbers. Functions defined on the natural numbers are very much a special case. Another special case is that of functions whose values lie on the two-element set

$$\text{Bool} = \{\mathbf{T}, \mathbf{F}\}$$

These are the *boolean-valued functions*[6].

There is no need to stop with pairs, or 2-tuples. Given the definition of the ordered pair, we can define the ordered n-tuple (for n > 2)

$$<x_1, x_2, \ldots, x_n> = << x_1, x_2, \ldots, x_{n-1}>, x_n>$$

We can associate an n-ary (n > 2) relation with a set of n-tuples. We can also associate an n-place function, a function which maps n arguments on to a value, with a set of (n + 1)-tuples. But we can see that, for n > 2, an n-tuple is really a set of pairs, since each n-tuple is a pair whose first element is an (n-1)-tuple and whose second is simply another element. So we can think of a function as a triple of domain, co-domain and set of ordered pairs, even in the case of an n-ary (n > 1) function. In that case the third component of the function is a set of (n+1)-tuples, and its domain is a set of n-tuples.

So an ordered n-tuple is actually a pair whose first element is an ordered (n-1)-tuple. Clearly, and in general, what an n-tuple is depends on what an (n-1)-tuple is. The apparent circularity is, however, not vicious because we start somewhere, in fact at n = 2, and then build up to any integer n, greater than 2. This definition of n-tuple is *recursive*. Recursive definitions, and the closely related category of inductive arguments together make up a topic in computer science whose importance is difficult to exaggerate. Before we turn to it, we briefly introduce the notion of the type of a function.

---

[6] **T** and **F** are the same as the two values **true** and **false** of the Pascal type boolean.

3.4 The type of a function

Consider the 2-place function **add**. **add** takes two natural numbers and delivers a third, its sum. So **add** is a set of 3-tuples

$$\{<0,0,0>, <0,1,1>,...<1,0,1>, <1,1,2>,....\}$$

or, more explicitly,

$$\{<<0,0>,0>, <<0,1>,1>,..., <<1,0>,1>, <<1,1>,2>,...\}$$

ordered pairs, each of which consists of

(1) a first element which belongs to $\mathbb{N} \times \mathbb{N}$, and

(2) a second element which belongs to $\mathbb{N}$.

We express this fact, that the domain of plus is $\mathbb{N} \times \mathbb{N}$ and its co-domain is N, by writing

$$\textbf{add}: \mathbb{N} \times \mathbb{N} \rightarrow \mathbb{N}$$

This characterizes the *type* of the function plus. Similarly, we write

$$\textbf{times}: \mathbb{N} \times \mathbb{N} \rightarrow \mathbb{N}$$

showing that **add** and **times** have the same type.

We can define a new function **new-add**, which is the set

$$\{<0, \{ <0,0>, <1,1>,...\}>,...., <1, \{ <0,1>, <1,2> \} >,....\}$$

This is a function, since it is a set of ordered pairs and since each argument is associated with only one value. But this time the value associated with each integer is a set of ordered pairs of integers. In fact each of the ordered pairs of integers is also a function. So the type of **new-add** is

$$\textbf{new-add}: \mathbb{N} \rightarrow (\mathbb{N} \rightarrow \mathbb{N})$$

The brackets '(' and ')' show that the co-domain of **new-add** is a function. **new-add** is a one-place function whose co-domain is another one-place function. Adopting the convention that '$\rightarrow$' is an operator that *associates to the right,*[7] we can drop the brackets, and write

$$\textbf{new-add}: \mathbb{N} \rightarrow \mathbb{N} \rightarrow \mathbb{N}$$

Clearly, plus and new-plus are equivalent in that if we supply them with the two arguments n and m, they both deliver the sum (n + m). However,

add 1 is undefined whereas **new-add** 1 has as its value a function, in fact, the function **successor.**

This device of taking an n-place function for (n > 2) and turning it into a one-place function whose co-domain is an (n - 1)-place function is called *currying* (after the American logician Haskell B Curry) the original function.[8]

[7] See chapter 5.

3.5 Recursive Definitions and Inductive Arguments

In section 3.2 we assumed that we already had the natural numbers $\mathbb{N}$ and the addition function '+', and we defined **successor** in terms of them. But it is instructive to work backwards and begin with the successor function, construct the natural numbers and the addition function.

Suppose we have **successor**[9]. Then we can define the natural numbers $\mathbb{N}$ recursively by the three conditions

(1) 0 is a natural number,
(2) if x is a natural number, then so is **successor**(x),
(3) there are no other natural numbers.

A recursive definition like the one above can appear to be circular. (2) already refers to 'natural number' in the antecedent part of the conditional. But our definition isn't viciously circular since it begins with an explicit non-recursively defined basis condition. The set that satisfies the *basis condition* is called simply the basis set. The basis condition (1) asserts that $0 \in \mathbb{N}$. A further condition, or conditions — in our case the single condition (2) — enables one to build more complex members of the set out of the simpler members. (2) tells us that $\mathbb{N}$ is *closed* under **successor** function. A final condition, our (3), tells us that an object that cannot be constructed by the previous conditions is not a member of the set we are defining. In other words, the recursively defined set is contained in all sets which satisfy the previous conditions. It is the smallest, or least such set.

Prolog programs consist largely of recursively defined boolean-valued functions, or predicates as we say later. Recall our first Prolog program

```
sum_to_N(0,0).

sum_to_N(s(N),S) :-
```

[8] It is a device we shall see applied in the construction of the semantics of programming languages, which we discuss in chapter 14.

[9] And assume we know some important properties possessed by **successor**, among which are that 0 is not the successor of any number and that a natural number is the successor of no more than one natural number.

```
    sum(N,S1),
    add(s(N),S1,S).

add(0,N,N).

add(s(N1),N2,s(N3)) :-

    add(N1,N2,N3).
```

and recall the function **add** on the natural numbers. The basis condition is

(addition-1) **add**(x, 0) = x

The closure condition is

(addition-2) **add**(x, s(y)) = s(add(x, y))

These two conditions appear in the Prolog program as two clauses which assert (i) that it is true that **0** and N and **N** stands in the 'add' relation, and (ii) that if **N1**, **N2**, and **N3** stand in the 'add' relation, so do **s(N1)**, **N2** and **s(N3)**.

How can you prove that something, say ψ, is true of every member of an infinite set like that of the natural numbers which has a first member and which is generated by a function like successor? You cannot examine them all in turn. But if you can prove that ψ is true of the first member, and that if ψ is true of a given natural number then ψ must be true of its successor, then you have proved that it must be true of every member of the set. That this is so is contained in *the principle of mathematical induction.* Schematically, given

(i) $\psi(0)$, and

(ii) for all $n \in \mathbb{N}$, if $\psi(n)$ implies $\psi($**successor**$(n))$

then we can infer that $\psi(n)$, for all $n \in \mathbb{N}$.

The appeal to induction affords probably the most powerful and most widely used method of proving facts about the structures — data types — that computing scientists and logicians have to deal with. Proof by induction naturally goes hand in hand with the recursive definition of data types, and we shall appeal to it later in the sketch proofs we give of the properties of logical systems.

The most common use of mathematical induction deals with induction on the natural numbers. The basis condition most often concerns zero, and the induction condition tells us that if every number up to n belongs to the set in question, then so does the number (n + 1).

Slightly more generally, one can replace the basis condition with a condition on some non-zero natural number, say k, and then prove by induction that every number greater than or equal to k is in the appropriate set.[10]

One of the most beautiful features of recursion is that it enables us to define interesting new data types elegantly and with a minimum of effort. Consider the cases of S-expressions, lists, strings and trees.

3.6 S-expressions

The ubiquitopus data type *List* is a special case of a very general structure called the *Symbolic-expression*, or *Sexpr* (pronounced 'S expression') for short. Begin with a basis of **atoms** and a 2-ary constructor function **mk-Sexpr** (read 'make S-expression') which generates a new Sexprs out of two Sexprs. Strictly, every choice of atoms generates a different type **Sexpr** (and **List**, etc.). We recursively define the set of Sexprs (that is **Sexpr**) over the domain of **atoms**[11] (or, to avoid tedium, plain 'Sexprs' for short). Call the set of Sexprs, **Sexpr**.

(1) an atom is an Sexpr (or belongs to **Sexpr**),
(2) if S_1 and S_2 are Sexprs,
then so is the pair **mk-Sexpr**(S_1,S_2),
(3) there are no other Sexprs.

Another name for pairing function **mk-Sexpr** is **cons**, for

[10] For example, suppose we want to prove that, according to our definition of ordered n-tuple

$$\text{if } \langle x_1,\dots,x_n\rangle = \langle y_1,\dots,y_n\rangle, \text{ then } x_1 = y_1,\dots, x_n = y_n.$$

then we can argue inductively as follows. We begin, not with n = 0, but with n = 2. Then for n = 2, the equality holds by definition of the ordered pair, which handles the basis condition. For the induction step, suppose that

$$\text{if } \langle x_1,\dots,x_m\rangle = \langle y_1,\dots,y_m\rangle, \text{ then } x_1 = y_1,\dots, x_m = y_m,$$

for $2 \leq m \leq n - 1$.
Now consider the next longest pair of tuples, when $\langle x_1,\dots,x_n\rangle = \langle y_1,\dots,y_n\rangle$.
Since this is really $\langle\langle x_1,\dots,x_{n-1}\rangle, x_n\rangle = \langle\langle y_1,\dots,y_{n-1}\rangle, y_n\rangle$ by definition, then since the first two ordered (n-1)-tuples must be equal, the induction hypothesis shows that $x_i = y_i$ for all $0 \leq i \leq (n-1)$, and finally, $x_n = y_n$ since the second elements of the ordered pairs must be equal by definition.

[11] We usually demand that **atoms** contain a special element called *nil*, which we write as [].

'constructor', as we noted. **cons** has the type

$$\mathbf{cons}: S \times S \rightarrow S$$

cons appears as one of the fundamental functions in the functional programming language LISP[12] and re-appears as a built-in function in Prolog. In Prolog we can represent the **cons** of S_1 and S_2 by either

$$.(S_1,S_2)$$

which mirrors our defintion exactly, or, and more usually, by

$$[S_1 \mid S_2]$$

The bar '|' represents the **cons** generator in an operator-like syntax, and its arguments are S_1 and S_2. Diagrammatically, we can think of it as a tree-like object:

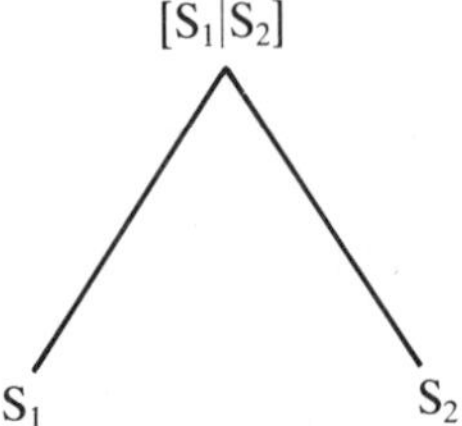

Assuming that **atoms** = { [] } ∪ {a, b, c, ... }, here is a diagram of the Sexpr cons(a, cons([], b)):

12 S-exprs are the sole data structure that LISP countenances. But LISP gets its name from the expression 'LISt Processing language', as any book on LISP will tell you. Lists are what LISP mostly deals with in practice, but lists are merely a special and especially convenient type of Sexprs.

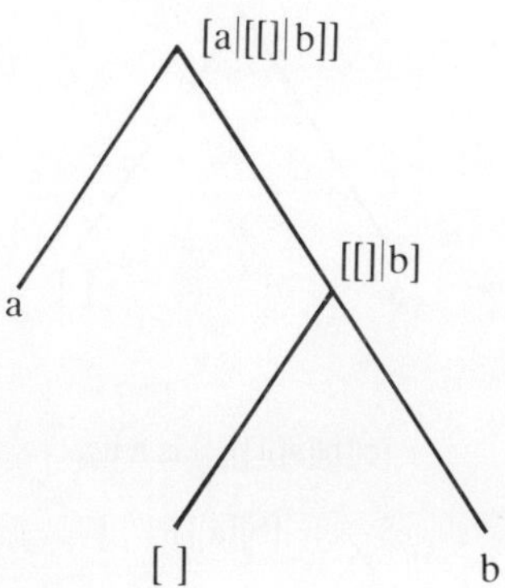

3.7 Lists and strings

Intuitively, lists are things like

$$[S_1, S_2, ... S_n]$$

where each is S_i either belongs to atoms, Sexprs . There are two points to note. First, **atoms**, Sexprs and Lists are all Sexprs. Now we demand that **atoms** contain a special element called nil, which we write as []. Secondly, the new constructor **mk-List** can again only form pairs. So how can we make Lists out of Sexprs? The solution is to represent any List as a pair, that is, as an Sexpr, but an Sexpr of a particular type.

Then we say that

(1) [] is a List (or belongs to **List**),

(2) if S_1 is an Sexpr and S_2 is a List, then so is the pair **mk-List**(S_1,S_2),

(3) there are no other Lists.

Clearly, [] is anomalously both an atom and the basic list. So if we begin with the set **atoms** = {[], a, b, c, ...}. The differences between Sexprs and Lists are (1) that any atom is an Sexpr whereas the only atomic list is [], and (2) that the second element in a pair which is a list must be a list, not just any Sexpr.

Examples

[] is a list
[a|[]] is a list

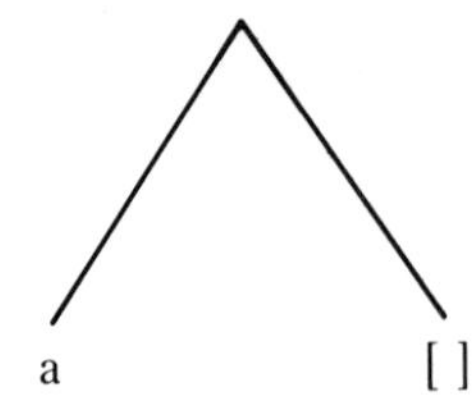

[c|[b|[a[|[]]]]] is a list

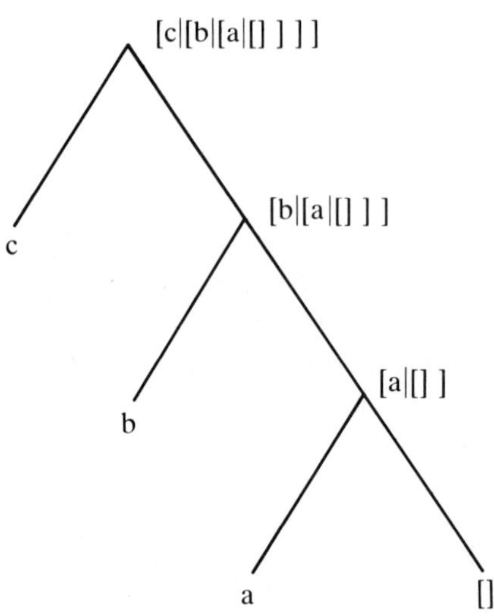

[[a|[]]|[]] is a list

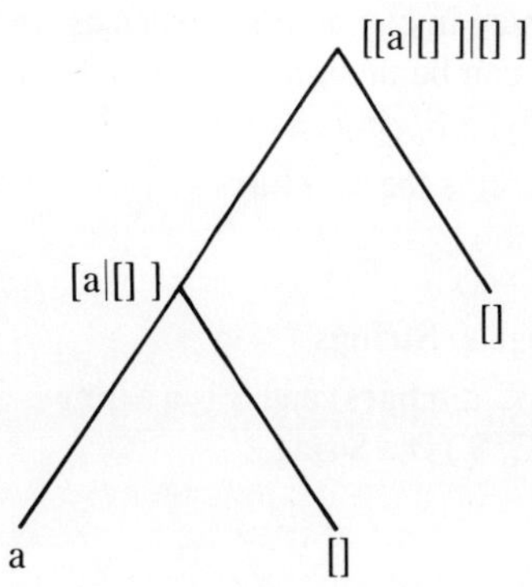

but [a|[[]|b]] is *not* a list.

Think of a list as a special kind of Sexpr in which [] appears as the 'right-most' Sexpr in the list. The generator **mk-List** for lists is again called cons, and is represented in Prolog as before. But strictly the two functions are different since they have different domains.

The type of the List generator is

$$\textbf{mk-List} : \textbf{Sexpr} \times \textbf{List} \rightarrow \textbf{List}$$

Whereas the type of the Sexpr generator **cons** is

$$\textbf{mk-Sexpr} : \textbf{Sexpr} \times \textbf{Sexpr} \rightarrow \textbf{Sexpr}$$

From now on, we adopt the Prolog notation. In the List [S|L] 'S' is called the *head,* and the List L is called the *tail.* [] has no head and no tail. The proliferating brackets cry out for a friendlier notation. So we represent the list

[a|[b|[c|[]]]]

by

[a, b, c].

[] is represented by itself, but when [] is the tail of a List it simply disappears in the representation, so that [a|[]] is represented as [a], which takes us back to our intuitive, but syntactically sugared, representation of a List.

To prove that every list has a property ψ we need not argue by mathematical induction on the length of lists. We can argue more directly. We show that [] has the property ψ, and that if a list L has ψ so does the list [S|L], where 'S' is any Sexpr. In fact, we can show by this *structural induction*[13] that every list has a length, a result we could not obtain by induction on the length of lists.

Lists turn up pretty well everywhere in computing science as we shall see. For example, character strings can be thought of as a type of List. More formally, one can define the type **Strings** as follows. Begin with a set **chars** of Characters. Our basis set is the set **chars** $\cup$ {[]}. Then we define

(1) [] is a String (that is, belongs to **Strings**)
(2) if C is a Character (that is, C $\in$ **chars**) and S is a String, then the pair **mk-String**(C, S) is a String,
(3) there are no other Strings.

The type of the String generator is

mk-String : Char × String → String

Compare the definition of Strings with that of Lists. Clearly a String, in our formulation, is a special kind of List. Non-nil Strings are Lists whose first element is a character (not just any Sexpr) and whose seond element is a string (not just any list).

So we can represent a string as a list of characters. This corresponds to what we do in Prolog, as we shall see. In Prolog a string is a List of ASCII codes.

3.8 Trees

Binary trees (Bintrees, the elements of the type **BinTree**) are most naturally represented as 'two-dimensional' objects. A binary tree is either [], or consists of a node and two binary trees (a left and a right binary tree, known as its sub-trees), . The basis set is **Node**, of which [] is not taken to be a member. By 'binary tree' we mean 'binary tree over the basis **Node**'.

(1) [] is a binary tree,
(2) if N is a Node, and if T_R and T_L are binary trees, then the pair **mk-Bintree** (N,T_L,T_R) is a binary tree,
(3) there are no other binary trees.

[13] Suppose that any object of a given type is either simple or is composed of parts, as a list is either [] or is composed of an Sexpr and a list. *If we can show that simple objects of the type always have a certain property* ψ *and that whenever the parts of an object of the type have* ψ *so does the object itself, then we can infer that all objects of the type have* ψ.

Here the generator is **mk-Bintree**. Its type is given by

$$\textbf{mk-Bintree: Node} \times \textbf{ BinTree} \times \textbf{BinTree} \rightarrow \textbf{BinTree}$$

The following jargon is pretty standard. The tree [] has no *children*, but a tree **mk-Bintree** (N,T_1,T_2) has T_1 and T_2 as its children, and it has N as its *root*. The nodes of a tree which have no children are called its leaves. In an ordered, binary tree, as we have chosen to represent it, only [] can be a leaf. Those nodes in the tree which are not leaves are called its internal nodes. [] has no root, no leaves, and no internal nodes.

We can represent the binary tree **mk-Bintree**(N,T_1,T_2) using a List. In fact we can choose one of several different representations like

$$[N,T_L,T_R]$$

or

$$[N, [T_L,T_R]]$$

Exercise 3.1

Generalize the idea of a binary tree to the idea of a tree in general (or an n-ary tree). How would you represent an n-ary tree using Lists ?

3.9 Discrete mathematics in Prolog

How can we represent sets, ordered pairs, relations, functions in Prolog?

We chose our notation for the **cons** generator for Lists to accord with the one employed by Prolog. Prolog uses [] for nil, as we did, and either the 2-ary functor '.' or the operator '|' for **cons**. Here for example, are the clauses for **occurs_in**/2, usually called **member**/2, but not to be confused with the membership relation of set theory.

```
/* occurs_in(Element,List) */

occurs_in(X,[X|_]).

occurs_in(X,[_|Tail]) :-
   occurs_in(X,Tail).
```

The clauses for **occurs_in/2** say that whatever is the head of a list belongs to it, otherwise, something belongs to a List if it belongs to its tail. Thus

| ?- *occurs_in(a,[a,b,c]).*
| ?- *occurs_in([a,b],[a,[a,b],c]).*

all succeed, whereas

| ?- *occurs_in([b,a],[a,[a,b],c]).*

doesn't. We know how to represent Lists in Prolog, they come pretty much as expected in the interpreter. But consider representing sets using lists. This leads to our first non-trivial Prolog problem. The sets {a,b}, {b,a} and {a,b,a} are the same set, a fact which tells that there is no single correct representation of a set. Yet we must represent sets in some way. In Prolog, it is most natural to represent a set as a list. Thus, we represent the set {a,b} by the list [a,b]. We can represent the set {b,a} by [b,a]. There is a problem here. Representing a set as a list should not imply that two lists represent the same set iff they are the same list. [a,b] is not the same list as [b,a], but both represent the same set. If we can't do anything about the fact that lists are ordered, we can rule out repetitions in our list representation of sets. We can adopt the convention that a list represents a set only if it contains no repetitions. Given a list, we can make a list which can then be said to represent a set. In Prolog we can define a predicate **mk_set/2** which is true of a list — List, say — and a list — Set, say — which represents a set. We can also define a predicate **is_a_set/1** which is true of a list which is an acceptable representation of a set. In the context of Prolog, we can say that a set is a list with no repetitions. Of course, a Prolog set is, strictly speaking, just an acceptable list-representation of a set. We can call such list representations of sets *Prolog-sets*. How can we define membership for Prolog-sets ? Here is one solution.

```
                    /* member(Element,Set) */

member(Element,[Head | Tail]) :-

    atomic(Element),
    !,
    (     Element = Head
          ;
          member(Element,Tail)   ).

member([H|T],[Head| Tail]) :-

    equal_sets([H|T],Head),
    !
    ;
    member([H|T],Tail).
```

Now we can define the fundamental relations between sets: 'equal_sets' and 'subset' — we have to, to make sense of 'member' — and one of the fundamental operations on a pair of sets: 'union'.

```
          /* subset(Set_1, Set_2) : Set_1 is a subset of Set_2

       union(Set_1,Set_2,Set_3) : Set_3 is the union of Set_1 and
                                  Set_2 */

sub_set([],_).

sub_set([Head | Tail],Set) :-

    member(Head,Set),
    !,
    sub_set(Tail,Set).
```

```
equal_sets(Set_1,Set_2) :-

    sub_set(Set_1,Set_2),
    !,
    sub_set(Set_2,Set_1).

union([],Set,Set).

union([Head | Tail],Set_2,Set) :-

    union(Tail,Set_2,Set_3),
    (   member(Head,Set_2),
        !,
        Set = Set_3
        ;
        Set = [Head | Set_3] ).
```

Exercise 3.2

Suppose one represented a set in Prolog as a List without repetitions.[14] Then [a,b] would be a set, but [a,b,a] would not be a set. Would [[a,b],[b,a]] be a set ?
Using this representation how would you define **member**/2?

Given our Prolog representation of sets we can define the predicate **is_an_ordered_pair**/3 as follows[15].

[14] See for example the section on *Lists and sets* in *Logic Programming and Knowledge Engineering,* Tor Amble, Addison-Wesley,1987. Are the definitions given there adequate, given the subtle differences between sets and lists ?

[15] Notice that according to our definition

$$\langle x,x \rangle = \{\{x\},\{x,x\}\} = \{\{x\},\{x\}\} = \{\{x\}\}$$

and this explains the first clause for **is_an_ordered_pair**/2.

```
                    /*ordered pairs*/

is_ordered_pair([[Element]],Element,Element).

is_ordered_pair([First,Second],X,Y) :-

   First = [X],
   !,
   (    Second = [X,Y],
        !
        ;
        Second = [Y,X])
   ;
   First = [Y],
   !,
   (    Second = [X,Y],
        !
        ;
        Second = [Y,X]).
```

Once we have ordered pairs we can develop the ideas of relations and functions. We can capture the notion of binary relation between two sets in Prolog as follows:

```
              /* binary_relation(R,A,B) :
        R is a two-place relation between A and B */

binary_relation([],_,_).

binary_relation([Ordered_Pair| Tail],Domain,CoDomain) :-

   is_ordered_pair(Ordered_Pair,X,Y),
   !,
   member(X,Domain),
   !,
   member(Y,CoDomain),
   !,
   binary_relation(Tail,Domain,CoDomain).
```

Note that we have assumed that the second and third terms of

binary_relation/3 are Prolog-sets.

Exercise 3.3

Write an 'efficient' version of 'union' which fails (as ours does not) if Set_1 and Set_2 are not Prolog-sets.

Exercise 3.4

The relative complement — A \ B — of two sets A and B is the set of all elements of A which do not belong to B. Give Prolog clauses for 'relative_complement(Set_1,Set_2)'.

Exercise 3.5

The powerset of a set is the set of all its subsets. The empty set belongs to the powerset of any set, since it is a subset of any set. If there are n elements of a set, then it has are 2n subsets. Write Prolog clauses for 'power_set(Set,Power_Set)' which generate a Prolog-set 'Power_Set' which the powerset of any Prolog-set 'Set'.

Exercise 3.6

Assuming a basis called *atomic_elements*, give Prolog clauses for

(1) **is_list**/1
(2) **is_string**/1
(3) **is_binary_tree**/1
(4) **is_tree**/1.

Summary

We outlined the basic mathematics needed for the semantics of logic — sets, ordered pairs, ordered n-tuples, relations, and functions. We showed how some of these mathematical ideas could be implemented in

Prolog. We also sketched the features of recursively defined data types like S-expressions, lists, trees, and strings.

4 Parsing with Prolog

Any language — whether a natural language like English, or formal like Pascal or Prolog — has *a syntax*, *a semantics,* and *a pragmatics.* The *syntax* of the language prescribes how the basic symbols of the language — the vocabulary **char** - may be strung together to form the strings of symbols that are the allowed sentences, or formulae of the language. Syntax omits all reference to meaning. We can handle the syntax of logic, and the syntax of a programming language, as if it were simply a game played with tokens, marks on paper, or characters on a screen. The *semantics* of the language, on the other hand, defines the meanings of the vocabulary and of the formulae of the language. Any formal language which interests us will have a semantics. Finally, and rather more vaguely, the *pragmatics* of the language is concerned with all aspects of the use of the language not captured by its syntax and semantics. When we speak of the pragmatics of a natural language, we are usually thinking the innumerable ways in which its actual use is constrained. In the case of programming languages, pragmatics is what is independent of its syntax and semantics, and dependent only upon its implementation.[1]

In talking *about* a language, whether formal or natural, we talk *in* a language. We call the language we talk about the *object language,* because it is the object of our discussion. In treating a formal language, the language in we talk about the object language in must be a different language, we call the *meta-language*. This discussion, indeed this book, is written in English — or rather, English laced with logic and a little mathematics — and so English is our meta-language.

Our object languages are principally the formal languages of formal

[1] Thus, though Pascal has a predeclared constant *maxint*, there is no particular maximum natural number defined by the semantics of Pascal. *maxint* varies from implementation to implementation.

logic. The distinction between object language and meta-language is intuitively quite clear. It is one which we must always be aware of if we are to avoid confusion, although maintaining the distinction with perfect strictness can be a surprisingly tricky business, a fact illustrated by what follows. If our object language is just a set of symbol strings — as it is both for logic and for a programming language — then we must be able to talk about particular strings in the language and about strings in the language in general. We will need constants to refer to fixed symbols and strings. And we will need variables in order to refer to any string. For example, suppose the basic vocabulary of our object language consists of the twenty-six lower-case alphabetic characters marked on a keyboard. Suppose the set of sentences, or allowed formulae, of the language is the set of lower-case French words with no accents. Then we can say:

'a' is in the basic vocabulary of our object language.

and

'ordinateur' is a sentence of the object language.

Both these sentences are in the meta-language. The first three marks in the first sentence (including the quotation marks) refer to the first character in the vocabulary. We *name* the first character in the vocabulary using those three marks. We also want to be able to say things like:

if S is in the basic vocabulary,
then S is one of the twenty-six lower-case alphabetic characters.

In this meta-linguistic sentence, the nineteenth upper-case letter of the English alphabet *ranges over* the basic vocabulary. It is a meta-linguistic variable, ranging over a domain. It is not to be confused with the nineteenth character in the basic vocabulary which is 's'. Nowhere in our discussion do we *use* symbols in the object language. We always *mention* them.[2]

Compare this with the similar case of English which, since it is a natural language, we must allow to be its own meta-language, subject to the use of a device like quotation to signal the mentioning of its own

[2] For an extended discussion of the distinction between use and mention see *Methods of Logic,* WVO Quine, p. 37ff, and also *Metalogic,* by Geoffrey Hunter, University of California Press, 1971, pp. 10-13.

expressions. We say

Logic is hard to do.
'Logic' is easy to spell.

We have to develop a formalism in which we can not only talk about, but can define object languages. A language is just a set of strings, so we have to find a way of specifying a formal language as a set of strings. In fact, we have to specify what strings are as well. The formalism we use for specifying a language as a set of strings is called the formalism of *phrase structure grammars.*

4.1 Phrase structure grammars

We begin, as in the last chapter, with a set of entities called **atoms**. Normally the set of our atoms contains the conventional typographical symbols. A string is a finite list (or *sequence)* of atoms. A string may be empty, in which case it contains no atoms and has length 0.

The fundamental operation on atoms and strings is provided by the **mk-String** constructor, which takes an atom and a string, and puts the atom at the *head* of the string, to make a new string. Another operation **concat** takes two strings and places the first in front of the second, again making a new string. We use capital letters, sometimes with subscripts as in S_1, to range over strings. We represent the concatenation of two strings S_1 and S_2 by **concat**(S_1,S_2), or more simply by merely juxtaposing the strings, as in

$$S_1S_2$$

Two strings are identical if either they are both empty, or consist of the same head and tail. A string S" is a sub-string of a string S' iff there are strings S_1 and S_2 (both or either may be empty) such that

$$S' = S_1S''S_2$$

A language may be just a set of strings of symbols chosen from some base set of atoms, but in a non-trivial language not all strings of atoms are going to acceptable sentences. We generally want to separate the sentences — those strings which are to count as 'grammatical' — from those strings which are not. The machinery we use for this task is a particularly useful one in computer science, in linguistics, and in formal logic: that of phrase structure grammars. A phrase structure grammar is a meta-linguistic structure which generates a language, the language generated being construed as a set of sentences. It consists of four items:

(1) a set of (meta-linguistic) variables which we allow to range over strings of the language, whether sentences or not;
(2) a set of (meta-linguistic) constants which designate the corresponding basic symbols, or 'terminals', of the language;
(3) a set of rules, called *production rules* which enable us to derive the sentences of the language; and
(4) a final special symbol, the so-called *start symbol,* for which the letter 'S' is usually reserved.

More formally, a phrase structure grammar is a 4-tuple

$$< S_N, S_T, P, S >$$

where
(1) S_N is a non-empty set of symbols — the set of meta-linguistic variables ranging over strings,
(2) S_T is a non-empty set of symbols — the set of (names of the symbols in the set) terminals (the atoms),
(3) P is a set (the *production set*) of production rules of the form

$$S_1 -> S_2$$

where S_1 is a non-empty string of terminals and non-terminals and S_2 is a (possibly nil) string of terminals and non-terminals — that is, both S_1 and S_2 may contain zero or more terminals and/or non-terminals, though S_1 must contain at least one terminal or non-terminal, and
(4) S is a variable designating the start symbol.

S_N , the set of 'non-terminals', and S_T, the set of 'terminals' must be disjoint, to avoid confusion between the 'variables' and the 'constants'.

Now we set up one method of specifying the sentences of a language. We proceed by the roundabout route of first specifying the *derivations* enjoyed by a phrase structure grammar — these lead to the *sentential forms* of the phrase structure grammar. The sentential forms are like sentences execpt that they can contain variables. Then we pick out those sentential forms which contain no variables and label these the sentences of the language we defined. First, the idea of a derivation. We can *directly derive* a string S_2 from a string S_1, which we write as

$$S_1 \to S_2$$

if and only if S_1 contains a substring which is the left side of one of the production rules of the phrase structure grammar and S_2 is S_1 with the right side of the production rule replacing that sub-string. Of course the language defined by the phrase structure grammar — the set of sentences of a phrase structure grammar — depends on its terminals but also, and more subtely and more interestingly, on its production rules. We can *derive* a string S_2 from a string S_1, written as

$$S_1 \Rightarrow S_2$$

if and only if there is a sequence of direct derivations in the phrase structure grammar from S_1 to S_2. Thus

$$S_1 \Rightarrow S_2$$

if and only if for some S_i, S_j, ...S_k,

$$S_1 \to S_i,$$
$$S_i \to S_j,$$
$$....$$
$$S_k \to S_2.$$

Example

Consider a very simple phrase structure grammar, called **simple**, for which

$$S_N = \{S\}$$
$$S_T = \{a,b\}$$

and whose production set is

$$P = \{\ S \to aSa,\ S \to b\ \}$$

Clearly the language generated by this grammar is the set

$$\{b,aba,aabaa,aaabaaa,.....\}$$ [3]

There is a very important hierarchy of four phrase structure grammar types, known as the *Chomsky Hierarchy,* which results from the placing of restrictions on the forms of the production rules that the phrase structure grammar may take. Here are the restrictions.

A phrase structure grammar is of type 3 if the left-hand side of each production rule is a single non-terminal (that is, the grammar is of type

---

[3] Or $\{a^n b a^n : n \geq 0\}$, where a^n means n 'a's in succession.

2), and if the right-hand side of each production rule contains at most one non-terminal symbol which must then be the right side symbol furthest to the right. A phrase structure grammar is of type 2 if the left side of each production rule is a single non-terminal symbol.

A phrase structure grammar is of type 1 if the length of the right-hand side of each production rule is at least as great as the length of its left-hand side.

All phrase structure grammar are trivially of type 0 by definition.[4]

Phrase structure grammars of types 2 and 3 are very important in computing science. The sets of Pascal and Prolog identifiers, for example, are generated by a phrase structure grammar of type 3. The sets of Pascal and Prolog expressions are generated by a phrase structure grammar of type 2. Languages generated by phrase structure grammars of types 2 and 3 are generally easy to parse efficiently, and parsers for them are relatively easy to implement in Prolog, as we shall soon see. There is special name for phrase structure grammars of types 2 (and hence for the special case of type 3 grammars): a phrase structure grammar of type 2 is called a *context-free grammar*. The language defined by a type 3 phrase structure grammar is said to be *regular*. We shall find that the language of the propositional calculus, the language of the predicate calculus, and even the syntax of Prolog itself are specifiable as context-free grammars.

4.2 BNF and EBNF

Context-free grammars are familiar to all computing scientists through the definition of large parts of the syntax of programming languages using grammars in *BNF*, or *Backus-Naur form*. BNF notation is simply a commonly-used variant representation of the production set of a context-free grammar. Non-terminals are written as identifiers surrounded by the brackets '<' and '>'; terminals are written as themselves, or as the names of sets of terminals; and the symbol '->' is written as '::='. Productions which a fixed non-terminal of the left are written as alternatives on the right using the symbol '|' which can be read as 'or'. As an example, consider the small language of numerals.

[4] From the definitions one can see that if a phrase structure grammar is of type n, then it is also of type m where $m < n$, (for $n,m \geq 0$).

Numerals

A numeral_string is either a digit, which is one of the terminals **0**,..**9**. Otherwise, it is a digit followed by a numeral_string. Thus **123** is a numeral_string because it is a digit **1**, followed by a **23**. **23** is a numeral_string because it is a digit **2** followed by **3**. And **3** is a numeral_string because it is a digit. Using N as the non-terminal standing for '<numeral_string>', and D as the non-teminal standing for 'digit', we have the productions

N	->	D
N	->	DN
D	->	**0**
D	->	**9**

The same production rules can be written in the more readable BNF as follows.

<numeral_string>	::=	<digit> | <digit><numeral_string>
<digit>	::=	**0** | ... | **9**

We compress the first two productions into the first BNF rule with two alternatives, and the next ten productions into the second BNF rule. But we can go further and extend the BNF formalism into EBNF, *extended* BNF. In EBNF one uses the curly brackets '{' and '}' to denote *zero or more iterations* of the string contained within. And one uses the square brackets '[' and ']' to denote *zero or one occurrences* of the string contained within. So we can write an EBNF grammar for numerals as

<numeral_string>	::=	<digit> {<digit> }
<digit>	::=	**0** | ... | **9**

The first rule

<numeral_string> ::= <digit> {<digit> }

is clearly equivalent to

<numeral_string> ::= <digit> | <numeral_string>

We can also write an EBNF grammar for numerals as

<numeral_string> ::= <digit> [<numeral_string>]
<digit> ::= **0** | ... | **9**

A second example is the language of identifiers.

Identifiers

We use both **numerals** and **identifiers** in writing an interpreter for a small programing language in chapter 14. So by an identifier we mean a string which is either a lower-case letter, or begins with a lower-case lettter and is followed by a string consisting of lower-case letters and digits. Using 'I' for identifier_string, 'L' for lower-case letter, 's' for a string consisting of lower-case letters and digits, and 'D' for digit we have the following unwieldy productions.

```
I  ->  L
I  ->  LS
S  ->  L
S  ->  D
S  ->  LS
S  ->  DS
L  ->  a
L  ->  z
D  ->  0
D  ->  9
```

In BNF we have a more elegant transcription of the grammar

<identifier_string> ::= <letter> | <letter><letter_digit_list>
<letter_digit_list> ::= <letter> | <digit> |
<letter><letter_digit_list>|
<digit><letter_digit_list>
<letter> ::= **a** | ... | **z**

<digit> ::= **0** | ... | **9**

And in EBNF we have the even more perspicuous

<identifier_string>	::=	<letter>	[<letter_digit_list>]
<letter_digit_list>	::=	<letter>	[<letter_digit_list>] |
		<digit>	[<letter_digit_list>]
<letter>	::=	**a** | ... | **z**	
<digit>	::=	**0** | ... | **9**	

Why should we bother with BNF and EBNF, if both are mere syntactic sugar for the productions of a context-free grammar ? The answer, for us, is that grammars written in BNF, and especially EBNF, forms may be easily transformed into parsers in Prolog, as we shall see.

4.3 The Prolog Grammar Rules

Prolog represents the string "abc" as the list of the corresponding ASCII codes [97,98,99]. The empty string, or list, is written as [], and is again pronounced 'nil'. So that "" is really [].

A beautiful feature of most implementations of Prolog are the *Prolog Grammar Rules.*[5] The Prolog Grammar Rules enable one to transcribe a context free grammar pretty directly into Prolog clauses which the interpreter allows to be represented in a special way. A pre-processor built into Prolog *translates* these special clauses, which do not conform to the syntax of ordinary Prolog, into genuine Prolog clauses which are then executable.We take the Prolog representation of **simple**, displayed in Fig. 4.1, as an example. This first Prolog program is an acceptor for strings of the language **simple**.

---

[5] Described for example in *Programming in Prolog,* W Clocksin and C Mellish, Springer-Verlag, 1982, Chapter 9.

```
                /* Prolog acceptor for simple
              using the  Prolog Grammar Rules  */


s --> a,s,a.          /* First Production */
s --> b.              /* Second Production */

a --> "a".            /* First Terminal */
b --> "b".            /* Second Terminal */
```

One can read the first 'production rule'

```
s --> a,s,a.          /* First Production */
```

as: *an 's' can be an 'a', followed by an 's', followed by an 'a'*. Thus, we read '-->' as "can be" and the commas as "followed by". This is not real Prolog of course, and so the Prolog pre-processor translates this program into the following:

```
s(X,Y) :-             /* First Production */
  a(X,X_1),
  s(X_1,X_2),
  a(X_2,Y).

s(X,Y) :-             /* Second Production */
  b(X,Y).

a(X,Y) :-             /* First Terminal */
  X = [97 | Y].

b(X,Y) :-             /* Second Terminal */
  X = [98 | Y].
```

Notice that the query in the comment at the beginning of the original acceptor shows **s** as having arity 2. Let's see how such a query as

| ?- *s("aba",Y).*

works, when issued to the real Prolog program into which the Prolog Grammar Rules are translated. Here is the commentary.

(1) | ?- **s("aba",Y).** is really | ?- **s([97,98,97],Y).**

(2) **s([97,98,97],Y)** succeeds by the First Production if the body succeeds, that is, if **a([97,98,97],X_1)**, and **s(X_1,X_2)**, and **a(X_2,Y)** all succeed.

(3) **a([97,98,97],X_1)** succeeds by the First Terminal, which it does for **X_1 = [98,97]**.

(4) **s(X_1,X_2)**, or in other words, **s([98,97],X_2)** succeeds by the Second Production, if **b([98,97],X_2)** succeeds.

(5) **b([98,97],X_2)** succeeds with **X_2 = [97]**, by the Second Terminal.

(6) **a([97],Y)** succeeds by the First Terminal, with **Y** = [].

One can represent this pictorially as follows. Notice that the call to **s([98,97],[97])** succeeds because of the second terminal production rule.

```
s( | 97 | 98 | 97 | , Y)   :-

a( | 97 | 98 | 97 | , | 98 | 97 | ),
s( | 98 | 97 | , | 97 | ),
a( | 97 | , [] ).
```

yes **Y** = []

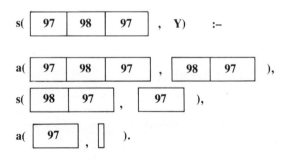

So "aba" is a string of **simple**, with trailing junk ""; in other words, no trailing junk. "aba" is a string of **simple**, full stop. This is our first quick sketch of how the Prolog Grammar Rules work. In the next chapter we use them to do more work. We use them to do some real parsing, for languages which are not quite so simple as **simple**.

4.4 Examples: numerals and identifiers

We can implement parsers for our languages **numerals** and **identifier** from chapter 3. These return — as Prolog atoms — the numeral and the identifier corresponding to the string defined by the production rules of their respective languages.

Note that in the following parsers some clauses have literals surrounded by curly brackets, as in **{name(N,N_String)}.** In the case of these literals the pre-processor's translating action is suppressed and the literal re-appears in the translated program in its original form of **name(N,N_String)**.

```
                    /* numerals */

numeral(N) -->

   numeral_string(N_String),
   {name(N,N_String)}.

/*
   <numeral_string>   ::=   <digit> [<numeral_string>]
   <digit>   ::=   0 | ... | 9
                                  */

numeral_string(N_String) -->

   digit([Digit]),
        (     numeral_string(N_String_1),
              {N_String = [Digit | N_String_1]}
              ;
              !,
              {N_String = [Digit]}      ).
```

```
digit([Digit]) -->

   [Digit],
   {48 =< Digit,
    Digit =< 57}.

        /* identifiers — strings of lower case letters and digits
                           beginning with
                        a lower-case letter */

identifier(I) -->

   identifier_string(I_String),
   {name(I,I_String)}.


/*
<identifier_string>     ::=  <letter> [<letter_digit_list>]
<letter_digit_list>     ::=  <letter> [<letter_digit_list>] |
                             <digit> [<letter_digit_list>]
<letter>                ::=  a | ... | z
<digit>                 ::=  0 | ... | 9  */

identifier_string(I_String) -->

   letter([Letter]),
        (    letter_or_digit_string(L_String_1),
             {I_String = [Letter | L_String_1]}
             ;
             !,
             {I_String = [Letter]}      ).
```

```
letter_or_digit_string(L_String)  -->

   (      digit([X])
          ;
          letter([X])      ),
          (    letter_or_digit_string(L_String_1),
               {L_String = [X | L_String_1]}
               ;
               !,
               {L_String = [X]}     ).
digit([Digit]) -->

   [Digit],
   {48 =< Digit,
    Digit =< 57}.

letter([Letter]) -->

   [Letter],
   {97 =< Letter,
    Letter =< 122}.
```

Exercise 4.1

Implement a parser for the following grammar for expressions:

```
<Exp>      ::=   NOT <Exp> |
                 - <Exp> |
                 <Exp_1> |
                 <Exp_1> = <Exp>

<Exp_1>    ::=   <Exp_2> |
                 <Exp_2> + <Exp_1> |
                 <Exp_2> - <Exp_1>

<Exp_2>    ::=   <Exp_3> |
                 <Exp_3> * <Exp_2>
```

<Exp_3> ::= <Id> | (<Exp>)

4.5 Another example: a simple language of sets

We have made much of the point that we must distinguish a set from our representations of it. We can define a context-free grammar which characterizes the language of strings which name sets of (and sets of sets of ... etc.) natural numbers. A string which names such a set — a set-string — begins with a left curly bracket '{', and ends with a right curly bracket '}'. In between comes a string which names a list of elements (separated by commas). The elements of a set may be also be sets, and so are named by sub-strings which are themselves set-strings.

Sets

<set> ::= { <list_of_elements> }

<list_of_elements> ::= empty | <numeral>| <numeral> ','
<list_of_elements> |
<set> | <set> ', '<list_of_elements>

Exercise 4.2

Write a parser for the simple language of *sets*.

Summary

We defined phrase structure grammars and considered the important special case of context-free grammars.

We showed how context-free grammars could be implemented in Prolog using the Prolog Grammar Rules.

5 Propositional calculus: syntax

Logic has traditionally been concerned with the study of valid arguments, with characterizing logical validity, with the study of what follows from what. Arguments occur in everyday life. But arguments occur in mathematics, in the form of proofs and derivations. In both cases a conclusion is derived from some premises.

There are many logics tailored to various purposes. For the present, and until chapter 15, our logic is *classical logic,* which is the logic of mathematics and hence the logic of computing science. Classical logic comes in two forms, a special form and a general form. The special, and restricted, form is called the *classical propositional calculus* — or just *the propositional calculus*.

The propositional calculus is a theory of logical validity at the crudest level, at the level at which logical validity is a result only of the way complete propositions are combined to form more complex propositions. The propositional calculus is our theory of logical validity as it appears at the level of purely propositional arguments. To the propositional calculus, propositions are opaque and their internal structure is invisible except to the extent that they are logical combinations of other propositions.

So why bother with a logical instrument as simple as the classical propositional calculus? Though crude, the propositional calculus is not entirely trivial. It exhibits many of the features of the more general predicate calculus, of which it is a proper part. And, besides modelling a small part of logical validity, the propositional calculus is a Boolean algebra, and has applications outside the strictly logical realm. For example, the propositional calculus can be used to help with simplifying logic circuits. But briefly, and most importantly, the propositional calculus forms a good introduction to the ideas that we must pursue more fully later, those *well-formed-formula, proof,* and *decision procedure*, and to imporant logical distinctions, like those between truth and validity, syntax and semantics, and those between theorems, logical truths and rules of inference.

The propositional calculus describes logical validity in purely propositional reasoning. What then is a proposition? A proposition is that which is expressed by the utterance[1] of a declarative sentence.

Arguments are composed of sentences which express propositions strung together in a certain way. Here, for example, is an argument expressed in English.

If there is inflation, unemployment will increase.
If the economy is not deflated, there will be inflation.
If the economy is deflated, unemployment will increase.
Therefore, unemployment will increase.

The *therefore* signifies a connection between the sentences which come before it — the premises — and what comes after it — the conclusion. The argument is simple enough, but perhaps just a little difficult to follow. It helps to work through it using an artificial symbolism.

5.1 How to symbolize propositional arguments

Let us use capital letters of the alphabet to represent propositions, as one uses letters for the variables in an algebraic equation. Ignoring the fact that the first three sentences in the argument consist of sub-sentences, we might represent the argument by:

A
B
C
Therefore D.

What is wrong with this? What is wrong is simply that this is too coarse a representation of the argument. We need to capture the

[1] We construe 'utter' in the broadest sense, encompassing speaking, writing, typing at a computer terminal and so on. Declarative sentences are sentences in a language, and they are uttered or written or transmitted in some way at some time and place, relative to all of which they are either true or false. We distinguish declarative sentences, from other sentences which express questions and commands, since these latter are neither true nor false.

relationship between the sentences in the argument. We need to capture the fact that the sentence

unemployment will increase

is a component of the first and second sentences in the argument, and is the fourth of the sentences in the argument. So let us try a richer representation. Let us use upper-case letters for the component sentences in the argument, and English words for what are intuitively our logical notions, and represent the argument by:

If I then U.
If not-D, then I.
If D then U.
Therefore U.

Going even further and representing 'if...then...' as in 'infix' binary operator ' $\Rightarrow$ ', 'not...' as a unary operator '$\neg$', and 'therefore' as the (meta-linguistic) symbol '$\vdash$', the argument can be represented by

$$\{I \Rightarrow U, \neg D \Rightarrow I, D \Rightarrow U\} \vdash U$$

or, *dropping the curly set-theoretic brackets*, and using the more usual P, Q and R for the componet sub-sentences of in the argument

$$P \Rightarrow Q, \neg R \Rightarrow P, R \Rightarrow Q \vdash Q$$

The three formulae on the left-hand side of the turnstile '$\vdash$' are called the *premises*, and the formula on the right is called the *conclusion* of the argument.

Exercise 5.1

Paraphrase the following natural language arguments into the language of the propositional calculus:

(a) *If we don't fire them, they're useless.*
If we fire them, they've failed.

Therefore, if they haven't failed, they're useless.

(b) *If we don't fire them, they've been successful.*
If we don't fire them, they seem to be useless.
Therefore, if they are successful, they seem to be useless.

(b) *Either Robin was a spy or he wasn't.*
If he was a spy, then they have all our secrets.
If he wasn't a spy, then we hanged him unnecessarily.
If we hanged him unnecessarily, Klaus will defect.
If Klaus defects then they have all our secrets, unless we kill him.
Therefore, either we kill him or they have all our secrets.

Which of these arguments are intuitively valid, which intuitively invalid?

5.2 Truth and validity

Perhaps we do not know what 'deflating an economy' is. But our notation can still help us to assess the argument. We might substitute other sentences for 'P', 'Q', and 'R' and convince ourselves that the conclusion follows from the premises. We can do this because the validity of an argument is a matter of its *form* and not of its *content*. It is a matter of the relations that hold between the sentences in the argument and not a matter of their content — what the sentences say.

For this example, maybe we just 'see' that the argument is valid. But not all examples are so easy. We should develop a full and complete theory — the classical propositional calculus — of this sort of propositional reasoning. The theory should begin by specifying the formal language of the propositional calculus. It should then present an interpretation, or semantics, for that language. It should develop various types of machinery for proving the validity of arguments. And then it should provide some *meta-theory* which describes the properties of the whole system (like its consistency for example).

We begin with some concepts. If logic deals with arguments and their validity or invalidity, what do we mean by argument validity? It is most important that validity not be confused with truth, nor form confused with content. Validity and invalidity are features of arguments. Truth and falsity are features of sentences. Truth and falsity are the truth-values of sentences. An argument is not a single statement.[2] It is made up of two things: first, a collection of premises and secondly, a conclusion. Call the premises and the conclusion the constituents of the argument. Each constituent, the conclusion and each of the premises, is a sentence. The validity or invalidity of an argument is a matter of its form. The truth or falsity of a sentence is a matter of its content and the way the world is and is only tenuously connected with how the world is. The connection between the validity or invalidity of an argument on the one hand, and the truth or falsity of its premises and conclusion on the other is therefore quite tenuous, as we suggested.

An argument is invalid if the premises can all be true and the conclusion false. Otherwise it is valid.

If we have a valid argument with a true conclusion we can infer nothing about the truth-values of the premises. If we have an invalid argument we can infer nothing about the truth-values of the premises from the truth-value of the conclusion, and nothing about the truth-value of the conclusion from the truth-values of the premises. Like our (valid) argument about inflation, the conclusion is only as reliable as that of the least reliable premise.

The connection between validity on the one hand, and truth and falsity on the other enables us to say this: *that if we have a valid argument we can infer that the conclusion is true if we know that all the premises are true, and we can infer that at least one of the premises is false if we know that the conclusion is false.*

We formalize propositional logic in order to make clear what it is that makes a propositionally valid argument valid. But we must begin with a purely syntactic characterization of the language of the propositional calculus. We must, in other words, view the language of the propositional calculus as a set of strings, marks on paper, patterns on a terminal screen, which have no interpretation or meaning at all, until we give them one.[3]

---

[2] It isn't easy to formulate this precisely. Our argument about unemployment could have been written

If there is inflation, unemployment will increase,and
if the economy is not deflated, there will be inflation, and
if the economy is deflated, unemployment will increase;
***therefore**, unemployment will increase.*

which is a single sentence though it does not, I think, comprise a single statement.

[3] In fact, we leave the formal semantics - the interpretation - of our language until

5.3 Specifying syntax

We begin with a basis of *propositional variables* or *atoms* — the terms are interchangeable. These are atomic propositions which we take as standing for propositions whose internal structure is of no concern to us. By convention, our atoms are "a,.., 'z', and 'A",...Z', (and if needed 'A_n',...,'Z_n' for each $n \in \mathbb{N}$).[4] We have a unlimited (actually denumerable) number of propositional variables available. They are the members of the set *atoms*.

We need shall use five logical connectives '$\neg$', '&', '$\vee$', ' $\Rightarrow$ ', '$\Leftrightarrow$' which stand for 'not...', ...and...',...or...', '...if...' and '...if and only if...' respectively. These operators will be placed between wffs, so we call this notation *infix* notation.[5] Actually, this is more than we strictly need.

We also need the brackets '(' and ')'. The atoms, the logical connectives and the brackets make up the vocabulary of the propositional calculus.

Finally, we need a grammar, a collection of formation rules which tell us which strings of items from the vocabulary belong to the set **pc_wffs** of well-formed-formulae (or *wffs* as they are called). Only wffs can represent English sentences and so can be constituents of our representations of valid arguments. So what are we take to be a wff?

Let the (meta-linguistic) variables α and β range over strings of symbols from the vocabulary of the propositional calculus, as they do in the picture below. Notice that α and β are meta-linguistic variables, they vary and range over objects which are the elements of a language. Neither the symbol α nor the symbol β is a wff, but each can be taken to *stand for* a wff. They stand outside the language we are developing, and range over and look down as the picture suggests.

Chapter 6. Given the semantics of Chapter 6, we can characterize logical validity in the propositional calculus. In Chapters 8 and 9 we give purely syntactic characterizations of validity.

4 When we implement parsers we use lower-case letters for our atoms. When we write proof we use upper-case letters.

5 As opposed to *Polish* notation, for which see the exercises.

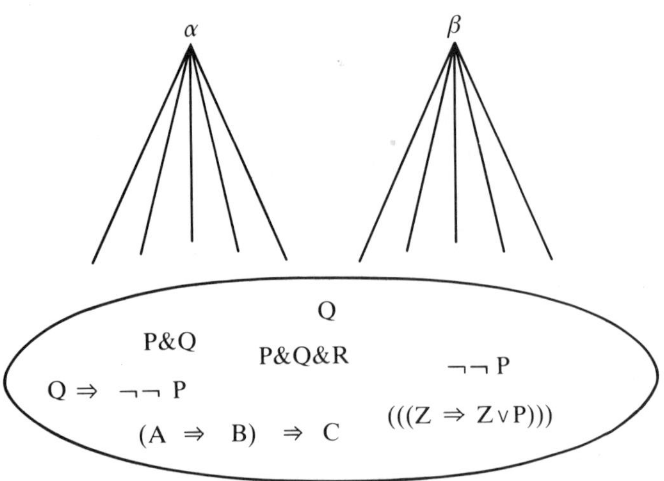

We begin with a definition that we shall quickly reject. To define wffhood in the propositional calculus it is usual to say:

(1) if α belongs to atoms, then α is a wff;

(2) if α is a wff, then so is $\neg\alpha$;

(3) if α and β are wffs, then so are

- (i) $(\alpha \,\&\, \beta)$;
- (ii) $(\alpha \vee \beta)$;
- (iii) $(\alpha \Rightarrow \beta)$; and
- (iv) $(\alpha \Leftrightarrow \beta)$;

(4) there are no other wffs

The brackets in (3) (i)-(iv) are meant to be taken seriously, at least at first. This is fine, except that it leads to an excess of brackets whenever there is an occurrence of one of the binary operators. Every occurrence of a connective, except '¬', is accompanied by a pair of brackets. There is a strong incentive to do something about the brackets. One solution is to associate a relative precedence with each connective.

How do we handle this? We say that the connectives '¬', '&', '∨', '⇒' and '⇔' are written in order of decreasing *precedence* — that '¬' has *higher precedence* than '&', and '&' than '$\vee$', and '$\vee$' than '$\Rightarrow$' and '$\Rightarrow$' than '$\Leftrightarrow$'. Or, in other words, and that '$\neg$' *binds more tightly* than '&' and so on. Connectives of higher precedence bind more tightly so that the string

$$\neg P \Rightarrow P \vee Q \Leftrightarrow R$$

is understood to mean

$$(((\neg P) \Rightarrow (P \vee Q)) \Leftrightarrow R)$$

much as in alegbra, where x∗y + y/z is understood to mean the same as (x∗y)+ (y/z). Strictly speaking of course, the first, abbreviated, bracketless string is not a wff at all, but is at best a conventional representation of the wff below it.

This business of introducing too many brackets then explaining them away is far too awkward, I think. The original simplicity of the definition of wffhood is completely lost. We allow a non-wff to be the proxy for a wff.

A better solution is to define the language **pc_wffs** as a context-free language from scratch, so that both fully bracketted and abbreviated strings count as wffs. The definition we give leads to a simple exercise in parser construction. There is, in any case, one issue which our previous discussion of wffhood left open. What, for example, are we to make of **P&Q&R** ? If we allow it as a wff (and why shouldn't we ?), should we think of it as **((P&Q)&R)** or as **(P&(Q&R))** ? Of course, it will turn out that the truth-values of the two possibilities are the same. The two distinct wffs are semantically equivalent, as we shall see. But this is not so for $\mathbf{P \Rightarrow Q \Rightarrow P}$ as we shall see We must distinguish semantically between $\mathbf{P \Rightarrow (Q \Rightarrow P)}$ and $\mathbf{(P \Rightarrow Q) \Rightarrow P}$. The former is a truth of logic, and the latter is not. In any case, here is an ambiguity we must clear up, the ambiguity being about the *associativity* of the binary connectives.

Binary connectives can be either *left-* or *right-associative*[6]. Consider the division operator '/ ' of ordinary algebra[7]. If we take

a / b / c

to mean the same as[8]

[6] Or indeed non-associative. For more on this see Section 5.6.

[7] We don't usually have to worry about associativities in algebra. When we do come across cases like a / b / c we put in unnecessary brackets, or re-write them, as either a / bc or ac/b.

[8] More strictly, ... *to have the same expression tree as...*

(a / b) / c

then '/' is *left*-associative. If we take it to mean

a / (b / c)

then it is *right*-associative. If we have a bracket-dropping convention such as the one above for **pc_wffs**, then we have to be explicit about the associativities of the binary connectives. Our context-free grammar for the propositional calculus contains its associativities implicitly. In fact in **pc_wffs** as we define it, all the binary connectives are *right*-associative.[9] Characterizing **pc_wffs** as a context-free grammar, we make the terminals of language of the propositional calculus the atoms, the brackets and the logical connectives. **pc_wffs** is then the class of strings defined by the following context-free grammar, in BNF form.[10]

<wff>	::=	<C_wff>	I	<C_wff> ⇔ <wff>
<C_wff>	::=	<D_wff>	I	<D_wff> ⇒ <C_wff>
<D_wff>	::=	<K_wff>	I	<K_wff> ∨ <D_wff>
<K_wff>	::=	<N_wff>	I	<N_wff> & <K_wff>
<N_wff>	::=	<factor>	I	¬ <N_wff>
<factor>	::=	<atom>	I	(<wff>)
<atom>	::=	a I ... I z I AI ...I Z		

Examples

P & Q ⇔ R ∨ S is a <wff> in virtue of the second condition for <wff>, <C_wff> ⇔ <wff>, and not in virtue of the first condition for <wff>, <C_wff> (and it will be a wff in virtue of one *or* the other and not both), and so the occurrence of '⇔' is its *dominant connective occurrence*.

Exercise 5.2

Define the expression **dominant connective occurrence** of a wff.

---

[9] This *is* a convention. Many logic text-books, when they consider the question at all, which is rare enough, make the binary connective left-associative.

[10] Much the same definition can be found in *The Science of Programming*, David Gries, Springer-Verlag, 1981, p.13.

Similarly, but less plainly ¬¬**(P & Q) & (Q ∨ R)** is a <wff> because it is a <C_wff>. It is a <C_wff> because it is a <D_wff>, because it is a <K_wff>, because it is of the form <N_wff> **&** <K_wff>, because ¬¬**(P & Q)** is an <N_wff> and **(Q ∨ R)** is a <K_wff>, and so its dominant connective occurrence is the second **&**.

5.4 Propositional Expression Trees

The fact that binary connectives appear in the propositional calculus leads us to associate a tree with each wff. The dominant connective occurrence is then the *root* of the associated tree, which we call the *expression tree of the wff.*

The *parse tree* of a wff of the propositional calculus is a tree whose structure represents the way in which the wff is derived, using the productions of the BNF grammar for the propositional calculus. More precisely, a parse tree represents a whole class of derivations. Thus, to give the parse tree of a wff is to describe it at a higher level of abstraction than that given by its derivation.

On the other hand, the expression tree of a wff represents the structure of the wff in an even more general way. Expression trees are a further abstraction of the real structure of a string — a mere sequence of characters. Take for example the two distinct wffs **(P & Q)** ⇔ ¬ **R** and **P & Q** ⇔ (¬**R**). Both of these have the same structure in the sense that the dominant connective of both is the '⇔'. The expression tree which represents this structure should intuitively have '⇔' as its root, and the expression trees of **P & Q** and ¬**R** as it left and right sub-trees. In this case, all the brackets are irrelevant. When they are not irrelevant, brackets serve merely to indicate the structure of the expression as represented by the expression tree, and are not part of the expression tree. But the brackets *are* part of the parse trees of the wffs. An expression tree throws away (or abstracts from) some of the information captured by a parse tree. Two different wffs may have the same expression tree, but they must have different parse trees. For example, the parse tree of the first wff is

```
S
├── C
│   └── D
│       └── K
│           └── N
│               └── F
│                   └── S
│                       ├── '('
│                       ├── C
│                       │   └── D
│                       │       └── K
│                       │           ├── N
│                       │           │   └── F
│                       │           │       └── A
│                       │           │           └── 'P'
│                       │           ├── '&'
│                       │           └── K
│                       │               └── N
│                       │                   └── F
│                       │                       └── A
│                       │                           └── 'Q'
│                       └── ')'
├── '↔'
└── S
    └── C
        └── D
            └── K
                └── N
                    ├── '¬'
                    └── F
                        └── A
                            └── 'R'
```

The parse tree for the second wff is

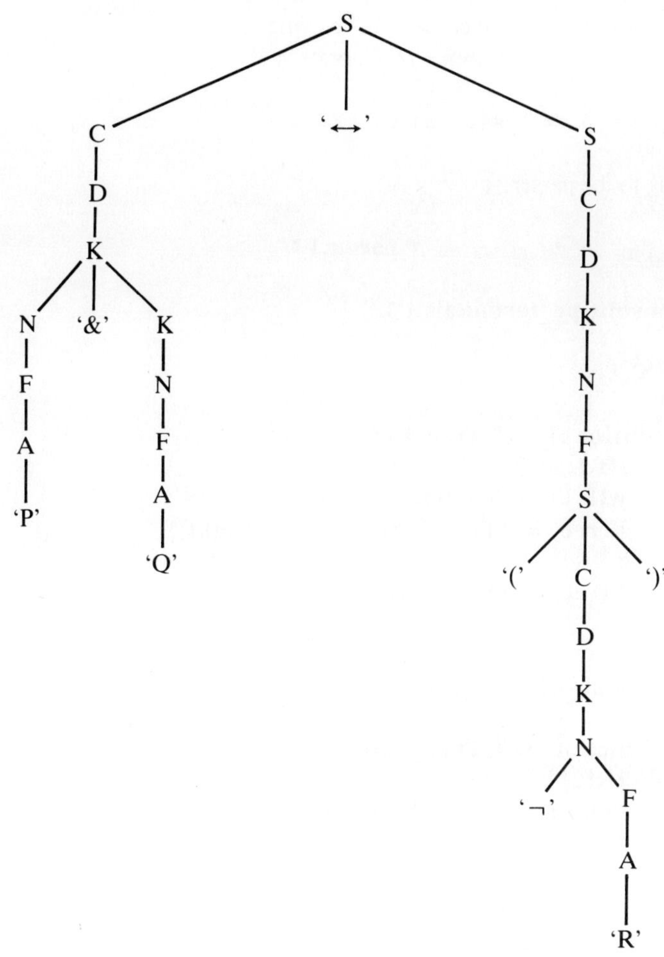

Can generate we parse trees automatically? Can we write a parser for **pc_wffs** which generates the parse tree of any wff of **pc_wffs**?

First, we re-write the BNF grammar for **pc_wffs** in EBNF form.

<wff>	::=	<C_wff> [	⇔	<wff>	]
<C_wff>	::=	<D_wff> [	⇒	<C_wff>	]
<D_wff>	::=	<K_wff> [	∨	<D_wff>	]

<K_wff> ::= <N_wff> [& <K_wff>]
<N_wff> ::= <factor> | ¬ <N_wff>
<factor> ::= (<wff>) | <atom>

<atom> ::= **a** | ... | **z** | **A**| ...| **Z**

leading to the parser

```
                    /* parser.1 */

:- consult('pc_terminals.1').

wff(['S',Tree]) -->

   conditional_wff(Tree_Left),
   (     iff,!,
         wff(Tree_Right),
         {Tree = [Tree_Left,'iff',Tree_Right]}
         ;
         {Tree = Tree_Left} ).


conditional_wff(['C',Tree]) -->

   disjunction_wff(Tree_Left),
   (     if,!,
         conditional_wff(Tree_Right),
         {Tree = [Tree_Left,'if',Tree_Right]}
         ;
         {Tree = Tree_Left} ).

disjunction_wff(['D',Tree]) -->

   conjunction_wff(Tree_Left),
   (     or,!,
         disjunction_wff(Tree_Right),
         {Tree = [Tree_Left,'or',Tree_Right]}
         ;
         {Tree = Tree_Left} ).
```

```
conjunction_wff(['K',Tree])  -->

    negation_wff(Tree_Left),
    (    and,!,
         conjunction_wff(Tree_Right),
         {Tree  =  [Tree_Left,'and',Tree_Right]}
         ;
         {Tree  =  Tree_Left} ).


negation_wff(['N',Tree])  -->

    negation_sign,!,
    negation_wff(Tree_Right),
    {Tree  =  ['neg',Tree_Right]}
    ;
    factor(Tree).


factor(['F',Tree])  -->

    left_bracket,!,
    wff(Tree_1),
    right_bracket,
    {Tree  =  ['(',Tree_1,')']}
    ;
    prop_var(Tree).
```

parser 1 consulted a file containing the terminals for the propositional calculus. This file is

```
                         /* pc_terminals.1 */

prop_var(['A',P_Var],[P|T],T)  :-

    P > 96,!,
    not(P = 118),!,        /* ascii code of 'v' */
    P < 123,!,
    name(P_Var,[P]).
```

```
iff --> "<->".
if --> "->".
or --> "v".
and --> "&".
negation_sign --> "-".
left_bracket --> "(".
right_bracket --> ")".
```

Here is the effect of the query

```
| ?- wff(Parse_Tree,"(p&q)<->-r",[]).
Parse_Tree =  [S,[[C,[D,[K,[N,[F,[(,
[S,[C,[D,[K,[[N,[F,[A,p]]],and,[K,[N,[F,[A,q]]]]]]]]],)]
]]]],iff,[S,[C,[D,[K,[N,[neg,[N,[F,[A,r]]]]]]]]]]
```

which is a very unpictorial representation of our first parse tree.

If we use a parse tree to represent the structure of a wff, we find that we are conveying redundant information. A parse tree tells us too much about how we derived the wff we derived. But we want to know what it represents. One piece of redundant information concerns the brackets. These tell us how to parse a wff but once the wff parsed, they carry no other information. An expression tree abstracts further than a parse tree and suppresses the history of the derivation of the wff — in other words the non-terminals — and the occurrences of the brackets '(', ')'.

Example

Here for example is the expression tree for the wffs

$$(\mathbf{P} \,\&\, \mathbf{Q}) \Leftrightarrow \neg \mathbf{R}$$

and

$$\mathbf{P} \,\&\, \mathbf{Q} \Leftrightarrow (\neg \mathbf{R}).$$

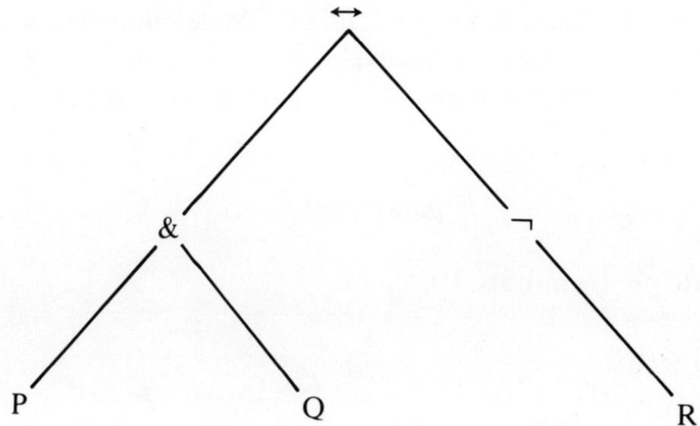

Given the parse tree of a wff, how can we generate its expression tree? These conditions describe the translation:

(1) If a wff's parse tree is **['A',Wff]**, then its expression tree is **Wff**.

(2) If a wff's parse tree is **[Non_Terminal,['neg',Sub_Parse_Tree]]**, then its expression tree is **['neg', Exp_Tree]** where **Exp_Tree** is the expression tree of **Sub_Parse_Tree**.

(3) If a wff's parse tree is **['Factor',['(',Sub_Parse_Tree,')']]**, then its expression tree is just the expression tree of **Sub_Parse_Tree**.

(4) If a wff's parse tree is **[Non_Terminal,[Left,Binary_operator,Right]]** then its expression tree is **[Binary_Operator,Left_1,Right_1]** where **Left_1** and **Right_1** are the expression trees of **Left** and **Right** respectively.

(5) Otherwise, the expression tree of **[Non_Terminal,Sub_Parse_Tree]**, is the expression tree of **Sub_Parse_Tree**.

These clauses exhaust all the possibilities. The definition lends itself to a Prolog implementation.Alternatively, one could re-write **parser.1** so that it delivers expression trees directly, rather than via parse trees. This gives us **parser.2**:

```
                         /* parser.2 */

:- consult('pc_terminals.2').

wff(Tree) -->

   conditional_wff(Tree_Left),
   (     iff,!,
         wff(Tree_Right),
         {Tree = iff(Tree_Left,Tree_Right)}
         ;
         {Tree = Tree_Left} ).


conditional_wff(Tree) -->

   disjunction_wff(Tree_Left),
   (     if,!,
         conditional_wff(Tree_Right),
         {Tree = if(Tree_Left,Tree_Right)}
         ;
         {Tree = Tree_Left} ).


disjunction_wff(Tree) -->

   conjunction_wff(Tree_Left),
   (     or,!,
         disjunction_wff(Tree_Right),
         {Tree = or(Tree_Left,Tree_Right)}
         ;
         {Tree = Tree_Left} ).
```

```
conjunction_wff(Tree) -->

   negation_wff(Tree_Left),
   (     and,!,
         conjunction_wff(Tree_Right),
         {Tree = and(Tree_Left,Tree_Right)}
         ;
         {Tree = Tree_Left} ).


negation_wff(Tree) -->
   negation_sign,!,
   negation_wff(Tree_Right),
   {Tree = neg(Tree_Right)}
   ;
   factor(Tree).


factor(Tree) -->
   left_bracket,!,
   wff(Tree),
   right_bracket
   ;
   prop_var(Tree).
```

Here are the new terminals.

```
                    /* pc_terminals 2 */

 prop_var(P_rep,[P|T],T) :-
   P > 96,!,
   not(P = 118),!,
   P < 123,!,
   name(P_rep,[P]).

 iff --> "<->".
 if --> "->".
 or --> "v".
 and --> "&".
```

```
negation_sign --> "-".
left_bracket --> "(".
right_bracket --> ")".
```

This time the query yields a more readable result.

```
| ?- wff(Tree,"(p&q)<->-r",[]).
Tree = iff(and(p,q),neg(r)) .

| ?- wff(Tree,"p&q<->(-r)",[]).
Tree = iff(and(p,q),neg(r))
```

5.5 A pretty-printer for wffs

Having built the expression tree of a wff, we can write a pretty-printer which re-writes the wff to the screen, but this time with the *minimum number of brackets*. If an expression tree of a wff is an atom, then the wff is an atom, and the pretty-printer should simply write the atom. So that

```
write_wff(Exp_Tree) :-

    atom(Exp_Tree),
    write(Exp_Tree).
```

But in all other cases — when either there are brackets or connectives, or both — the pretty-printer has to consider whether or not to write brackets. To write the minimum number of brackets, it is simplest first to assign precedences to each wff, depending on its dominant connective (if any):

```
precedence(and(_,_),4).
precedence(or(_,_),3).
precedence(if(_,_),2).
precedence(iff(_,_),1).
precedence(_,5).
```

One wants to write brackets around a sub-wff if the precedence of the root of that sub-wff is numerically less than the precedence the root of the wff immediately above it in the expression tree. We count atoms and negations as having precedence 5, as shown by the final clause for **precedence/2**. So here are two typical clauses, one governing negation, the other governing the conditional. The remaining clauses are easy to write.

```
write_wff(neg(Exp_Tree))  :-
   precedence(Exp_Tree,N),
   write('-'),
   (     N < 5,
         !,
         (     write('('),
               write_wff(Exp_Tree),
               write(')') )
         ;
         write_wff(Exp_Tree)      ).
```

We want to write brackets around the negated sub-wff if and only if the negated sub-wff is neither an atom, nor a negation. In other words, we want to write brackets around the negated sub-wff if and only if the precedence of the negated sub-wff is less than 5 (the precedence of every wff which does not have a dominant binary connective).

```
write_wff(if(Exp_Tree_1,Exp_Tree_2))  :-
   precedence(Exp_Tree_1,N_1),
   precedence(Exp_Tree_2,N_2),
   (     N_1 < 2,
         !,
               (     write('('),
                     write_wff(Exp_Tree_1),
                     write(')') )
               ;
               write_wff(Exp_Tree_1) ),
   write(' -> '),
   (     N_2 < 2,
         !,
               (     write('('),
                     write_wff(Exp_Tree_2),
                     write(')') )
               ;
               write_wff(Exp_Tree_2) ).
```

In the case of the conditional we must check the precedences of both sub-wffs and write brackets around the left sub-wff, and the right sub-wff, only when either has an '⇔' as its dominant binary connective.[11]

Exercise 5.3

Complete the pretty-printer which pretty-prints the propositional expression trees generated by parser.2.

5.6 Using operators

Compound terms are represented in Prolog as a functor, followed by a left bracket, followed by a sequence of terms separated by commas, followed by a right bracket, as in *Functor(.....,Term,...)*. But functors can also be treated as operators, in which case this convention can be sidestepped. For example, '+' is an infix operator. The query

| ?- *X* = *1* + 2, *Y* =.. *X*.
succeeds with

X = 1+2
Y = [+, 1, 2]

Prolog allows us to declare our own operators using the built-in **op**/3 predicate. **op**/3 has three places for the operator's Precedence, Type, and Name, in the order **op(Precedence, Type, Name)**. Operator precedence is represented by a natural number. For example, the built-in binary arithmetic operators of '+' and '∗' have, in C-Prolog, the following pre-declarations.[12]

```
:- op(500, yfx, '+').
:- op(400, yfx,'*').
```

---

[11] Continue the example at Exercise 5.3.

[12] Note: Operators with higher precedence have a lower precedence number. The numbers appear to vary from one implementation to another.

The **Type** of an operator tells us whether or not it is prefix, postfix or infix, and if it is infix what its associativity is. Prefix operators are represented by the type **fx** or **fy**. **fx** is used if the operand — the term operated on, in this case on the right — always has lower priority, otherwise the more general **fy** is used. Postfix operators are represented by the type **xf** or **yf**. **xf** is used if the operand — in this case on the left — always has lower priority, otherwise the more general **yf** is used.

The most interesting cases are those of the infix operators. Here we use the following:

xfx for non-associative infix operators. (In this case both the operands, on the left and on the right, must be of higher precedence — that is *lower* precedence number)
xfy for **right**-associative infix operators (the left operands must be of higher precedence)
yfx for **left**-associative infix operators (the left operands must be of higher precedence)

So, to declare a right-associative conjunction and disjunction on might insert into one's program code

```
:- op(600, yfx, '&').
:- op(590, yfx,'v').
```

Using the appropriate declarations[13] one can read a wff (followed by the term terminating full-stop) with

```
| ?- read(X).

|p & q <-> (-r).

X = <->(&(p,q),-r))
```

This is a cheap and cheerful method of implementing term-readers, but it has the disadvantages that one needs to put a full-stop at the end of a term, and that one learns relatively little about parsing. You need to know more about parsing to handle the syntax of the predicate calculus and the toy programming language of chapter 14.

---

[13] See Exercise 5.8.

5.7 Sequents

The aim of logic is to separate not merely logically true formulae from those that are not, but to separate the valid from the invalid arguments. Tautologies are wffs, and so correspond to sentences. Arguments lead from a set of premises to a conclusion. They are best thought of as an ordered pair whose first element is a set of wffs, the premises, and whose second element is a wff, the conclusion. We represent them as sequents.

We write sequents in the following way $\Gamma \vdash \alpha$ where Γ is a set (possibly empty) of wffs, and 'α' a wff. The '$\vdash$' can be read as 'implies'. Note that it is a meta-linguistic symbol and is therefore quite different from the conditional connective ' $\Rightarrow$ '.

$\{\alpha\} \vdash \alpha$ is, strictly speaking, a sequent. We drop the curly brackets around the premises, and write $\alpha \vdash \alpha$. It says " the set whose member is 'α' implies 'α' ".

$\alpha \Rightarrow \alpha$, by contrast, is a wff.

A sequent $\Gamma \vdash \alpha$ is *valid* iff there is no assignment of truth-values to the propositional variables which occur in its wffs which makes the conclusion α false and all the premises Γ true.

Two of the styles of logical proofs that we later handle, sequent calculi (in passing) and natural deduction systems (in detail), take sequents as their fundamental objects. In treating the propositional calculus as a formal system of logic it is often more convenient to think of sequents as the objects which we prove and to think of provable 'wffs', which turn out to be the tautologies, as a special kind of sequent.[14]

Exercise 5.4

Given a BNF definition of <sequent> and implement a parser for sequents in Prolog.

[14] Tautologies correspond to a special case of valid sequents, in which the premises is the empty set.

Exercise 5.5

Given the following BNF grammar for WFFS, write in Prolog a parser which supplies the expression tree of any wff of the language.

<wff> ::= <atom> | (¬ <wff>) | (<wff><bin_op><wff>)
<bin_op> ::= **<->** | **->** | **v** | **&**

Exercise 5.6

In Polish notation one represents wffs as strings in the following way.

(i) An atom is represented as itself.

(ii) A negated wff ¬<wff> is represented by the string

$$N\alpha$$

where α is the Polish representation of <wff>.

(iii) A wff

$$\text{<wff1>} \Leftrightarrow \text{<wff\_2>}$$

whose dominant connective occurrence is the binary connective $\Leftrightarrow$ as shown is represented by

$$E\alpha\beta$$

where α is the Polish representation of <wff_1> and β is the Polish representation of <wff_2>. (C is used for $\Rightarrow$, K for &, and D for $\vee$).
Write a parser for pc_wffs which generates the Polish representation for a wff. Write a parser which accepts wffs in their Polish form and outputs the infix form with the least number of brackets.

Exercise 5.7

Implement clauses for **exp_tree** such that **exp_tree** succeeds with the parse tree of a wff yielding its expression tree.

Exercise 5.8

Give operator declarations and write clauses for **analyse**/2 which result in the following session:

```
| ?- analyse(p,X).
X = p .
yes
```

```
| ?- analyse(p & ( q v r ) <-> ( p & q) v (p & r),X).
X = [<->,[&,p,[v,q,r]],[v,[&,p,q],[&,p,r]]] .
yes
| ?- analyse(p & q <-> - (- p & - q),X).
X = [<->,[&,p,q],[-,[&,[-,p],[-,q]]]] .
yes
| ?- analyse( p -> (- - - - - - p) v q,X).
X = [->,p,[v,[-,[-,[-,[-,[-,[-,p]]]]]],q]] .
yes
```

Summary

We argued that we can get a better understanding of the form of an argument if we represent it in a formally defined symbolism. Truth and validity must be sharply distinguished. Truth applies to sentences, or propositions; validity to arguments. We defined the formal syntax of the propositonal syntax and implemented the formal EBNF definition as a parser written in Prolog. We defined parse trees and expression trees for propositional calculus wffs. We introduced the idea of sequents. The propositional calculus, as a formal theory, is a theory which which deals with, and explains, the validity sequents.

6 Propositional calculus: semantics

The language of the propositional calculus is, thus far, simply a formal language whose elements are strings of symbols specified by a vocabulary (a set of terminal symbols), and formation rules given as a context-free grammar. We are free to impose on it any consistent interpretation that we choose. We choose to regard it as the language of a simple formal logic in which the propositional variables take as values the two truth-values **T** and **F**.

An assignment of truth-values to the propositional variables then induces an assignment of truth-values to all the wffs, as we shall see. Wffs which are not propositional variables are built up from propositional variables, the logical connectives (and the brackets). The truth-values of wffs which are not propositional variables are functions of the truth-values of their constituent propositional variables. This is what we mean when we say that the propositional calculus is a truth-functional propositional logic.

A binary connective, like '&' — which is a terminal of the language of the propositional calculus — has a corresponding syntactic function $\&_{\mathbf{syntactic}}$. Given two wffs, $\&_{\mathbf{syntactic}}$ conjoins them to form a third. So that we can write

$$\&_{\mathbf{syntactic}} : \text{Wffs} \times \text{Wffs} \rightarrow \text{Wffs}$$

Crudely, for wffs **P** and **Q**

$$\&_{\mathbf{syntactic}}(\mathbf{P}, \mathbf{Q}) = \mathbf{P} \,\&\, \mathbf{Q}$$

But '&' also has a corresponding semantic function $\&_{\mathbf{semantic}}$. The truth-value of a conjunction depends only on the truth-values of the two conjuncts. If this were not the case we could not say that $\&_{\mathbf{semantic}}$ is a function. If $\&_{\mathbf{semantic}}$ were not a function, our logic would not be truth-functional. Conjunction clearly is truth-functional, so corresponding to **&** we have the function

$$\&_{\mathbf{semantic}} : \text{Bool} \times \text{Bool} \rightarrow \text{Bool}$$

Bool is the set {**T**, **F**}. Wffs is an infinite set. So $\&_{\mathbf{semantic}}$ is a much smaller and simpler function than $\&_{\mathbf{syntactic}}$. In fact,

$$\&_{\mathbf{semantic}} = \{<T,T,T>,<T,F,F>,<F,T,F>,<F,F,F>\}$$

6.1 The truth-tables

We can best display $\&_{\mathbf{semantic}}$ as a table — the truth-table for conjunction. Here we assume that **P** and **Q** are meta-linguistic variables ranging over propositional variables of the propositional calculus. (Elsewhere we use 'p', 'q', 'r',... or 'A', 'B', 'C',...as meta-linguistic variables, as well as α, β, γ,...)

P	**Q**	**P**	**&**	**Q**
T	**T**		**T**	
T	**F**		**F**	
F	**T**		**F**	
F	**F**		**F**	

Similar remarks go for disjunction '∨' and its corresponding functions $\vee_{\mathbf{syntactic}}$ and $\vee_{\mathbf{semantic}}$. In this case

$$\vee_{\mathbf{syntactic}} : \text{Wffs} \times \text{Wffs} \rightarrow \text{Wffs}$$

and

$$\vee_{\mathbf{semantic}} : \text{Bool} \times \text{Bool} \rightarrow \text{Bool}$$

but

$$\vee_{\mathbf{semantic}} = \{<T,T,T>,<T,F,T>,<F,T,T>,<F,F,F>\}$$

so that the truth-table for disjunction is

P	Q	P	∨	Q
T	T		T	
T	F		T	
F	T		T	
F	F		F	

For the sole unary connective '¬' we have

$$\neg_{\textbf{syntactic}} : \text{Wffs} \rightarrow \text{Wffs}$$

and

$$\neg_{\textbf{semantic}} : \text{Bool} \rightarrow \text{Bool}$$

but

$$\neg_{\textbf{semantic}} = \{\langle T,F\rangle,\langle F,T\rangle\}$$

which is best represented as the truth-table

P	¬P
T	F
F	T

It so happens that the truth-functionality of '¬', '&' and '∨' as well as the actual truth-tables, we have given for them are pretty well uncontroversial. But the claim that '⇒' and '⇔' are truth-functional, and the truth-tables for '⇒' and '⇔' that we shall give, need more justification and explanation. The truth-tables for '⇒' and '⇔' are more difficult to swallow than those for '&' and '∨'. It can seem a mistake to think of '⇒' and '⇔' as truth-functional at all. Nevertheless, here are the incomplete truth-tables, as far as our intuitions take us:

P	Q	P	⇒	Q	P	⇔	Q
T	T		?			?	
T	F		F			F	
F	T		?			F	
F	F		?			?	

A conditional will surely be said to be false if its antecedent is true and its consequent is false. An equivalence is surely false if the two 'equivalents' are of opposite truth-values. But what about the other places left undecided with question-marks? We adopt a *convention*. It is a convention that simplifies our logic and yields a logic which is perfectly adequate to capture the kind of reasoning we do in mathematics and the kind we would ideally employ in extracting information from a database. The convention is: fill in the question marks with **T**. This leads to the following truth-tables for '⇒' and '⇔'.

P	Q	P ⇒ Q	P ⇔ Q
T	**T**	**T**	**T**
T	**F**	**F**	**F**
F	**T**	**T**	**F**
F	**F**	**T**	**T**

Clearly, it is easier to say when a conditional or an equivalence is definitely false, than when it is definitely true. The convention we adopt is *when in doubt, let them be true*. We said that an assignment of truth-values to the propositional variables induces an assignments of truth-values to all the wffs. Given all the propositional variables **atoms**, a *valuation* is defined to be a function of the type

valuation : **atoms** → Bool

If there are n propositional variables, then there are 2^n different valuations. Each propositional variable may be assigned **T** or **F**, independently of the assignments to other propositional variables. Intuitively, each propositional variable is independent of all the others because it is not 'made up' of any of the others. That is what makes it an 'atom'. So any assignment of truth-values to the propositional variables is admissible. But having made an assignment of truth-values to the propositional variables, we fix the values that we can assign to all the other wffs.

The truth-tables tell us that a valuation, which merely assigns truth-values to the propositional variables, may be uniquely extended to an *admissible_valuation* which assigns a truth-value to every wff. An admissible_valuation is a function of type

admissible_valuation : Wff → Bool

That a valuation extends uniquely to an admissible_valuation can be proved simply by structural induction on the size of wffs. The ideas of *valuation* and *admissible_valuation* enable us to say more about what the truth-tables represent. The top line of the truth-table for conjunction

P	Q	P	&	Q
T	T		T	

not only tells us that the conjunction of two true wffs is another true wff. For given propositional variables **P** and **Q**, it also *represents* a whole set of valuations and hence a whole set of admissible_valuations. It says that in every valuation in which **P** is true and in which **Q** is true, every corresponding admissible_valuation makes **P&Q** true.

Remember that in writing the truth-tables we assumed that **P** and **Q** were meta-linguistic variables ranging over propositional variables of the propositional calculus. Now we can extend the idea of truth-tables from propositional variables to wffs in general

Since each valuation extends to an admissible_valuation, we can think of the top row of the truth-table for conjunction as asserting that for any wffs **P** and **Q**, if both **P** and **Q** are assigned t in an admissible valuation, then so is **P&Q**.

6.2 Tautologies, contradictions, contingencies

Semantically, we distinguish between three kinds of wffs in the propositional calculus.

A *tautology* (or a *valid* wff) is a wff that always takes the truth-value **T** for all admissible valuations.

An *inconsistency* is a wff that always takes the truth-value **F** for all admissible valuations.

A *contingency* is a wff that takes the truth-value **T** for at least one admissible valuation, and the truth-value **F** for at least one admissible valuation.

We also say that a formula is *satisfiable* if and only if it takes the truth-value **T** for at least one admissible valuation. And we say that is *unsatisfiable* if and only if it does not take the truth-value **T** even for one admissible valuation. So a wff is an inconsistency if and only if it is unsatisfiable. It is a contingency if and only if neither it nor its negation is unsatisfiable. It is a tautology if and only if its negation is unsatisfiable

To determine whether wff is a tautology, an inconsistency or a contingency we can write out its truth-table. The connectives have their truth-tables and, by extension, so does each wff. The truth-table for a propositional variable like 'P' is trivial. It consists of the two rows which capture its two possible truth-values.

P
T
F

Every propositional variable is therefore a contingency precisely because for every admissible in which it is true there is one in which it is false.

Of course, there are plenty of tautologies containing occurrences of just one propositional variable. Here are three produced by the program sketched in Section 6.3. In that program '⇔' and '⇒' are represented by '<->' and '->' respectively, because the terminal screen cannot write them. '¬' is represented by '-', here re-edited to appear as '¬'.

```
p          p v ¬p
T          T T FT
F          F T TF
```

```
p          p -> p
T          T T  T
F          F T  F
```

```
p          ¬(p & ¬p)
T          T T F  FT
F          T F F  TF
```

The 'important' tautologies are the simple tautologies. Here are some important tautologies. At least two of them have names. 'p v ¬p' has the name *Law of Excluded Middle*. '¬(p & ¬p)' has the name *Principle of Non-Contradiction*. To assert 'p ∨ ¬p' is to assert that something equivalent to "p is either true or false, whatever proposition 'p' is". Similarly, to assert '¬(p & ¬p)' is to assert something equivalent to "it is not the case that both 'p' and '¬p' are both true". Notice that in the truth-tables it is the truth-value assigned to the dominant connective (hence 'dominant') that is the truth-value of the wff determined by the truth-values of its component propositional variables.

Take two wffs A and B. Collect together all the propositional variables that occur in A or in B or in both. If A takes the truth-value T whenever B takes the truth-value T for every assignment of truth-values to those propositional variables then A and B are said to be logically equivalent. If A and B are logically equivalent then A $\Leftrightarrow$ B is a tautology.

Some logical equivalences are so important that they are given titles. They are 'laws of logic'. Thus $\neg\neg A \Leftrightarrow A$ is known as the 'law of double-negation'. Its truth-table is

p	¬¬p	<->	p
T	TFT	T	T
F	FTF	T	F

$$A \,\&\, (B \vee C) \Leftrightarrow (A \,\&\, B) \vee (A \,\&\, C)$$

and

$$A \vee (B \,\&\, C) \Leftrightarrow (A \vee B) \,\&\, (A \vee C)$$

are known as the *distributive laws*. Their truth-tables are

a	b	c	a	&	(b	v	c)	<->	a	&	b	v	a	&	c
T	T	T	T	T	T	T	T	T	T	T	T	T	T	T	T
T	T	F	T	T	T	T	F	T	T	T	T	T	T	F	F
T	F	T	T	T	F	T	T	T	T	F	F	T	T	T	T
T	F	F	T	F	F	F	F	T	T	F	F	F	T	F	F
F	T	T	F	F	T	T	T	T	F	F	T	F	F	F	T
F	T	F	F	F	T	T	F	T	F	F	T	F	F	F	F
F	F	T	F	F	F	T	T	T	F	F	F	F	F	F	T
F	F	F	F	F	F	F	F	T	F	F	F	F	F	F	F

and

a	b	c	a	v	b	&	c	<->	(a	v	b)	&	(a	v	c)
T	T	T	T	T	T	T	T	T	T	T	T	T	T	T	T
T	T	F	T	T	T	F	F	T	T	T	T	T	T	T	F
T	F	T	T	T	F	F	T	T	T	T	F	T	T	T	T
T	F	F	T	T	F	F	F	T	T	T	F	T	T	T	F
F	T	T	F	T	T	T	T	T	F	T	T	T	F	T	T
F	T	F	F	F	T	F	F	T	F	T	T	F	F	F	F
F	F	T	F	F	F	F	T	T	F	F	F	F	F	T	T
F	F	F	F	F	F	F	F	T	F	F	F	F	F	F	F

$A \& B \Leftrightarrow \neg(\neg A \vee \neg B)$ and $A \vee B \Leftrightarrow \neg(\neg A \& \neg B)$ are known as 'de Morgan's Laws'. Their truth-tables are

a	b	a	&	b	<->	¬(	¬a	v	¬b)
T	T	T	T	T	T	T	FT	F	FT
T	F	T	F	F	T	F	FT	T	TF
F	T	F	F	T	T	F	TF	T	FT
F	F	F	F	F	T	F	TF	T	TF

a	b	a	v	b	<->	¬(	¬a	&	¬b)
T	T	T	T	T	T	T	FT	F	FT
T	F	T	T	F	T	T	FT	F	TF
F	T	F	T	T	T	T	TF	F	FT
F	F	F	F	F	T	F	TF	T	TF

Finally, recall our argument about unemployment. We can exhibit its validity by conjoining its premises, and making that conjunction the antecedent of a conditional whose consequent is the conclusion of the argument:

```
p q r    (p -> q) & (¬r -> p) & (r -> q) -> q
T T T     T T  T  T  FT T  T  T  T T  T  T  T
T T F     T T  T  T  TF T  T  T  F T  T  T  T
T F T     T F  F  F  FT T  T  F  T F  F  T  F
T F F     T F  F  F  TF T  T  T  F T  F  T  F
F T T     F T  T  T  FT T  F  T  T T  T  T  T
F T F     F T  T  F  TF F  F  F  F T  T  T  T
F F T     F T  F  F  FT T  F  F  T F  F  T  F
F F F     F T  F  F  TF F  F  F  F T  F  T  F
```

Alternatively (and equivalently), we could consider just those rows of the truth-table in which `(p -> q)`, `(¬r -> p)` and `(r -> q)`, are true and see whether `q` is true in those rows.

If we do not want to mirror the structure of English sentences in a direct way, we can dispense with '⇒' and '⇔'. That is, we can eliminate every occurrence of '⇒', '⇔' and 'v' from a wff using the equivalences

$$\mathbf{P \Leftrightarrow Q} \text{ is equivalent to } \mathbf{\neg(P \,\&\, \neg Q) \,\&\, \neg(\neg P \,\&\, Q)}$$

$$\mathbf{P \Rightarrow Q} \text{ is equivalent to } \mathbf{\neg(P \,\&\, \neg Q)}$$

$$\mathbf{P \vee Q} \text{ is equivalent to } \mathbf{\neg(\neg P \,\&\, \neg Q)}$$

A corresponding set of equivalences can be used to eliminate all occurrences of '⇔', '⇒' and '∨'.

Exercise 6.1

Write our the truth-tables for the following wffs. Which are tautologies, contradictions, contingencies?

(a) $\neg P \Rightarrow (P \Rightarrow Q)$

(b) $(P \,\&\, Q) \vee (\neg P \,\&\, \neg Q)$

(c) $(P \Rightarrow (Q \Rightarrow R)) \Leftrightarrow (P\&Q \Rightarrow R)$

(d) $\neg(P \Rightarrow (Q \Rightarrow P))$

6.3 Printing out truth-tables

A small case-study in Prolog programming, and in reading Prolog code.

The truth-tables listed earlier in this chapter were output from the Prolog program sketched here. The functor **table**/1 takes a string which is a wff of the propositional calculus, generates an expression tree, determines the list of atoms in the wff, writes the top line of the truth — the atoms in the wff and the wff itself, and then writes the rest of the truth-table. Hence

```
table(Wff) :-

    wff(Exp_Tree,Wff,[]),
    list_of_atoms(Exp_Tree,Atom_List),
    write_top_line(Atom_List,Exp_Tree),
    write_table(Atom_List,Exp_Tree),
    nl.
```

wff/3 is as in chapter 5. **list_of_atoms**/2 operates by taking an expression tree and making the list of the atoms contained in it, by examining cases. Notice how the clauses work by examining atoms, negations and wffs with a dominant binary connective.

```
list_of_atoms(Exp_Tree,List) :-

    make_list(Exp_Tree,[],List).
```

Here is how **make_list**/3 might be coded.

```
make_list(Exp_Tree,List_1,List_2) :-

   atom(Exp_Tree),
   !,
   add_to_list(Exp_Tree,List_1,List_2)
   ;
   Exp_Tree = neg(Exp_Tree_1),
   !,
   make_list(Exp_Tree_1,List_1,List_2)
   ;
   (    (    Exp_Tree = iff(Exp_Tree_1,Exp_Tree_2)
             ;
             Exp_Tree = if(Exp_Tree_1,Exp_Tree_2)
             ;
             Exp_Tree = or(Exp_Tree_1,Exp_Tree_2)
             ;
             Exp_Tree  =  and(Exp_Tree_1,Exp_Tree_2)
   ),
        make_list(Exp_Tree_1,List_1,List_3),
        make_list(Exp_Tree_2,List_3,List_2)   ).
```

Exercise 6.2

Write Prolog clauses for **add_to_list**/3 which adds a propositional variable to a list of propositional variables in the correct alphabetical order but only if the propositional variable is not already in the list.

So that

add_to_list(p,[],X)	succeeds with [p]
add_to_list(r,[p,q,r,s],X)	succeeds with [p,q,r,s]
add_to_list(q,[p,r],X)	succeeds with [p,q,r]

etc.

---

write_top_line/2 simply displays the appropriate terms

```
write_top_line(Atom_List,Exp_Tree) :-

    tab(2),
    write_atom_list(Atom_List),
    tab(8),
    write_wff(Exp_Tree).
```

write_table/2 does all the real work. Notice how **write_table**/2 works by iteration to prevent too large an environment building upon the stack:

```
write_table(Atom_List,Exp_Tree) :-

    initialise(Atom_List,IV),
    asserta(val(IV)),
    nl,
    repeat,

          val(V),
          write_line(Exp_Tree,V,Atom_List),
          next_valuation(V,NV),
          retract(val(_)),
          asserta(val(NV)),

    all_false(NV),
    !,
    write_line(Exp_Tree,NV,Atom_List),
    nl.
```

initialise/2 creates a list of **true**'s each corresponding to a value of a propositional variable in Atom_list. It sets up the truth-values of the propositional variables for the top line of the truth-table.

```
initialise(Atom_List,Initial_Val) :-

   Atom_List = [],
   !,
    Initial_Val = []
   ;
   Atom_List = [Head | Tail],
   initialise(Tail,Tail_1),
   Initial_Val = [true | Tail_1].
```

Within **write_table/2**, **write_line/2** and **eval/4** do the most work:

```
write_line(Exp_Tree,V,Atom_List) :-

        tab(2),
        write_truth_values_1(V),
        tab(8),
        eval(Exp_Tree,V,Atom_List,Val_Tree),
        !,
        write_row(Val_Tree),
        nl.

write_truth_values_1([]) :- !.

write_truth_values_1([Head | Tail]) :-

   (    Head = true,
        write('T ')
        ;
        Head = false,
        write('F ')     ),
        write_truth_values_1(Tail).
```

Here are the clauses for **next_valuation/2**.

```
next_valuation([true],[false]).
```

```
next_valuation([Head | Tail], Valuation) :-

   (      Head = true,
          all_false(Tail),
          !,
          initialise(Tail,Tail_1),
          Valuation = [false | Tail_1]      )
   ;
   (      next_valuation(Tail,Tail_1),
          Valuation = [Head | Tail_1]      ).

all_false(Valuation) :-

   Valuation = [],
   !
   ;
   Valuation = [Head | Tail],
   !,
   Head = false,
   all_false(Tail).
```

Notice how **eval**/4 works by cases. Here are the clauses dealing with the **Exp_Tree** of an atom, a negation, and a conditional.

```
eval(Exp_Tree,Val,Atom_List,Val_Tree) :-

   atom(Exp_Tree),
   assign(Exp_Tree,Val,Atom_List,Value),
   Val_Tree = Value.

eval(Exp_Tree,Val,Atom_List,Val_Tree) :-

   Exp_Tree = neg(Exp_Tree_1),
   eval(Exp_Tree_1,Val,Atom_List,Val_Tree_1),
   Val_Tree = neg(Val_Tree_1).
```

```
eval(Exp_Tree,Val,Atom_List,Val_Tree) :-

   Exp_Tree = if(Exp_Tree_1,Exp_Tree_2),
   eval(Exp_Tree_1,Val,Atom_List,Val_Tree_1),
   eval(Exp_Tree_2,Val,Atom_List,Val_Tree_2),
   Val_Tree = if(Val_Tree_1,Val_Tree_2).
```

Given the current valuation, we want to assign the appropriate value to a propositional variable

```
assign(Atom,[Head_1 | Tail_1], [Head_2 | Tail_2],Value) :-

   Atom = Head_2,!, Value = Head_1
   ;
   assign(Atom,Tail_1,Tail_2,Value).
```

Exercise 6.3

Write the Prolog code for **evaluate**/3. Simplify your code using **=..** .

6.4 Truth-functional completeness

It may be surprising that every truth-function can be represented by restricting the connectives to the set {'¬','&'}. Similarly, one can get by with the set {'¬','∨'}, and with the set {'¬', '⇒'}, but not with the set {'¬', '⇔'}.

The first three sets are functionally complete sets of connectives, and so we can translate every wff — semantically speaking, *every truth-function* — into a logically equivalent wff, or*equal truth-function*, which contains occurrences of only the two connectives in any one of the three functionally complete sets listed above. The proof is by induction on the number n of places of the truth-function f. So let f have the type

$$f : \text{Bool}^n \rightarrow \text{Bool}$$

We consider the set {'¬','&'}. The other cases follow similarly.

First, the number of rows in the truth-table for the function f is 2^n.

If n = 0, f is a constant, so there is one row. If n = 1, there are two rows. If n = 2 there are four rows, and so on.

If n = 0, there are, as we said, only two truth-functions — the constants whose values are **T** and **F**. So, representing our truth-function as the interpretation of a wff , the latter might be ¬(**P** & ¬**P**) for the first truth-function, and (**P** & ¬**P**) for the second, for some fixed atom **P**.

On the other hand, if n ≠ 0, then write out the truth-table of f. The first line of the truth-table assigns all the v_i's **T**. So it is equivalent to

$$v_1 \& v_2 \& \dots \& v_{n-1} \& v_n$$

The second row assigns **F** to v_n , so it is equivalent to

$$v_1 \& v_2 \& \dots \& v_{n-1} \& \neg v_n$$

While the last, and 2^nth row is equivalent to

$$\neg v_1 \& \neg v_2 \& \dots \& \neg v_{n-1} \& \neg v_n$$

So now we have 2^n of these conjunctions, each one containing a unique combination of negated and un-negated variables.

The trick is then to pick out the rows to which f assigns **T**, and take their disjunction. This yields a wff of the form

$$C_i \vee C_j \vee \dots \vee C_k$$

where $1 \leq i,j,\dots,k$, where each of the C's is a conjunction of the above form, and where the overall disjunction is truth-functionally equivalent to f. Now using de Morgan's Laws we can eliminate the disjunctions since

$$C_i \vee C_j \vee \dots \vee C_k \Leftrightarrow \neg(\neg C_i \& \neg C_j \& \dots \& \neg C_k)$$

The formula on the right is equivalent to f and contains only occurrences of only '¬' and '&'.

For example, suppose we have the following truth-function f of three arguments (call the arguments 'p', 'q', and 'r').

p	q	r	f(p,q,r)
T	T	T	T
T	T	F	F
T	F	T	F
T	F	F	T
F	T	T	F
F	T	F	F
F	F	T	F
F	F	F	T

f(p,q,r) takes the value **T** only in the first, fourth and last rows. So f(p,q,r) is equivalent to

```
  (p & q & r)
v (p & ¬q & ¬r)
v (¬p & ¬q & ¬r)
```

which, according to de Morgan's Laws, is equivalent to

```
¬(¬ (p & q & r)
  & (p & ¬q & ¬r)
  & ¬ (¬p & ¬q & ¬r) )
```

A similar argument shows that {'v', '¬'} is truth-functionally complete. Since we can define '&' in terms of '¬' and '⇒', it follows that {'¬', '⇒'} is also truth-functionally complete.

Exercise 6.4

Propositional logic — or more precisely Boolean algebra — can be used for simplifying logic circuits. We can represent an inverter — a *not* gate — and an *and* gate by the following symbols.

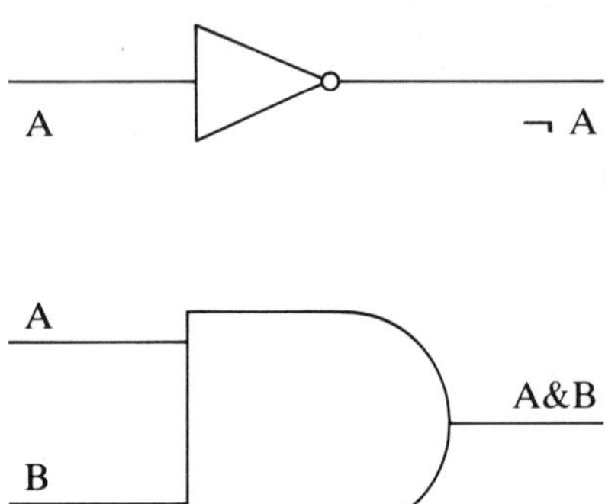

The inputs A and B can be thought of **T** and **F**, for 'on' or 'high', and 'off' or 'low'. How would you represent this circuit in the propositional calculus and to what can you simplify it?

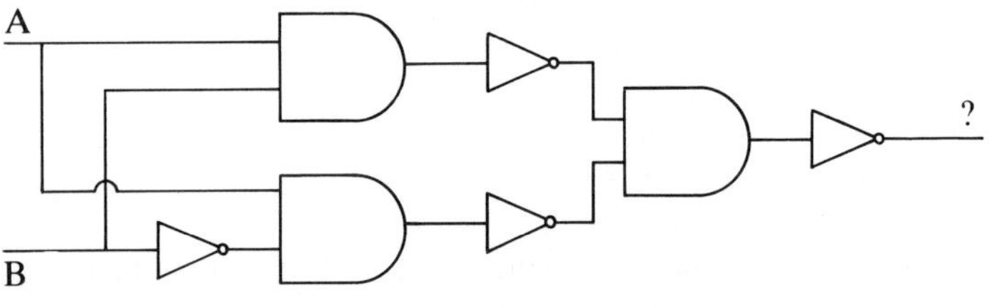

Summary

The atoms, or propositional variables, of the propositional calculus are interpreted as denoting for the truth-values, **T** and **F**. The meanings of the connectives are given by the truth-tables using propositional variables. Every truth-function can be represented by a truth-table. And every truth-table is the truth-table of a wff whose only connectives are '¬' and '&'.

A tautology is a wff that always takes the truth-value **T** for all admissible valuations. An inconsistency is a wff that always takes the truth-value **F** for all admissible valuations. A contingency is a wff that takes the truth-value **T** for at least one admissible valuation, and the truth-value **F** for at least one admissible valuation. We can test for tautologyhood, inconsistency, and contingency using the truth-tables. We saw have to write a Prolog program which writes out the truth-table for any input wff.

7 Four styles of theorem-proving

We are now in a position to check whether a given string is a wff of the propositional calculus. If a string is a wff of the propositional calculus then the truth-tables for the logical connectives enable us to determine whether it is a tautology, a contingency or a contradiction. But we want to do more than *check* whether a wff is a tautology. We want some machinery which will enable us to *prove* a wff valid in a sequence of steps.

The machinery which allows us to prove wffs is a luxury when designed for the propositional calculus, because the truth-tables allow us to do everything we want in the way of testing for tautologyhood and validity. But when we come to the predicate calculus we shall find nothing corresponding to the truth-table test for validity. In fact, in the predicate calculus, we have no option but to resort to proofs. So we examine theorem-proving in the simpler environment of the propositional calculus first.

7.1 Decision procedures, soundness, completeness, proof procedures, refutation procedures

We need some definitions, and we need to discuss what significance the definitions have. A *decision procedure* for some property ψ of a set of objects is an algorithm that determines whether or not any element of that set has ψ. The truth-table test gives us a decision procedure for tautologyhood (and for contingency and for contradictoriness) for the set of wffs of the propositional calculus. It also gives us a decision procedure for validity of sequents of the propositional calculus. Given a sequent — a pair whose first element is a set of wffs, the premises, and whose second element is a wff, the conclusion — we can use the truth-tables to check whether or not it is valid: we check that for all assignments of truth-values to the propositional variables which occur in the sequent, whenever all the wffs in the premises are true, at least one of the wff in the conclusion is true.

Tautologyhood and validity are a matter of semantics, a matter of the interpretation we give to the language of the propositional calculus. But though the syntax we choose for a formal language will naturally be guided by an intended interpretation, a formal language can, in the first instance, be viewed as a purely syntactic entity.

We therefore ask: is there a purely syntactic method of characterizing those wffs which turn out to be tautologies? is there a purely syntactic method of characterizing those sequents which are similarly valid?

The answer to both of these questions is that there is. In fact, there are many different purely syntactic methods of capturing the set of tautologies and the set of valid sequents. Like the rules which specify the formal language of the propositional calculus, each such method can be regarded as a game played with wffs (plus a few more incident symbols perhaps), a game which is meaningless until we give it a meaning.

A purely syntactic system which captures the tautologies will, by definition, contain no reference to truth and falsity, or to the meanings of the logical connectives. So we cannot regard it straight off as a system which captures tautologies or valid sequents. We must regard it as a system which captures a certain class of syntactic objects, certain wffs or certain sequents, which just happen to be those wffs which are tautologies or valid sequents This talk of 'capturing certain wffs and sequents' needs to be sharpened up. We should say that we mean to constuct one or more proof systems for the propositional calculus. A proof system is a game played with symbols according to strict rules, a game which has no meaning until we give it one. A proof system for the propositional calculus results in theorems. These are wffs or sequents which terminate proofs of the system.

A proof is, in general, a syntactic structure, usually a sequence or a tree of syntactic objects — a sequence or a tree of wffs or sequents (or, speaking very generally, perhaps sets of wffs, and so on) in which each item in the structure stands in a certain relation to earlier objects in the structure. The 'certain relation' is usually specified by a set of constraints called *rules of inference*.

One's intention in setting up a proof system for the propositional calculus is that a wff will be provable within it (and so be a theorem of it) if and only if it is a tautology, and a sequent provable if and only if it is valid. But, to repeat, such semantical notions can play no part in the proof system itself. A proof is just structure built out of syntactic objects, a structure which hangs together by satisfying certain rules.

Given that we have two different properties of a wff or a sequent — the semantic property of validity and the proof-theoretic property of provability — we should ask how they hang together. The answer is that in a properly constructed logic — that is, in a logic whose proof system is correctly constructed given a prior semantics, or a logic whose semantics is correctly constructed given a prior proof system — the two properties will coincide.

A proof system for a formal logic is *sound* for an interpretation if every provable wff or sequent is valid according to the interpretation.

A proof system for a formal logic is *complete* with respect to an interpretation if every wff or sequent which is valid according to the interpretation is provable.

All the proof systems which we discuss in this book are sound and complete with respect to their intended interpretations. Proving soundness and completeness is a fundamental activity for the logician, who is generally less interested in theorems within a logic (like de Morgan's Laws) than in mathematical results *about* a logical system. Logicians are interested in *meta*-logic. Proving soundness, and more especially completeness, for a given proof system of a given logic is a meta-logical activity and can be difficult. In this book we skate over such problems, since our interest, unlike that of the real logician, is with logic in practice, in fact, in the practice of computing and in software engineering.

If one has a proof system which is sound and complete with respect to an interpretation for which one has a decision procedure, then clearly one has a decision procedure for provability. But one can reasonably ask whether or not one has a procedure for *generating* the proof of a valid wff. Such a procedure, if one has one, is called a *proof procedure*, a procedure for generating proofs of provable wffs or sequents.

A procedure which is both a decision procedure and a proof procedure is called a *proof/decision procedure*.

So what kinds of proof system for the propositional calculus are there? We consider four kinds: the axiomatic, natural deduction, sequent calculus, and resolution theorem-proving kinds.

7.2 Axiomatic systems, natural deduction systems, and sequent calculi

Of these four kinds of proof system, the axiomatic systems are the oldest. Their history goes back to the nineteenth century, in fact to Frege's monumental work published in 1879 which we mentioned in passing in chapter 1. Bertrand Russell and Alfred North Whitehead's system set out in their *Principia Mathematica* of 1910 is an axiomatic system.[1] The German mathematicians Hilbert and Ackermann.[2] developed one in the 1920's, since when axiomatic systems have tended to be called 'Hilbert-Ackermann systems'.

Both natural deduction systems and sequent calculi, on the other hand, are usually considered to be the inventions of Gerhard Gentzen,[3] who published versions for the propositional and predicate calculi, and for some other systems as well, in both these new logical styles in a classic paper of 1935.

Finally, resolution theorem-proving, the subject of the next chapter, is the culmination of some of the researches of the 1950s and 1960s into *automatic* theorem-proving in general. Resolution theorem-proving embodies a proof procedure for the propositional calculus, is therefore perfected suited to handling by computer, and, in a fuller, predicate calculus version, forms the basis of the inference engine built into Prolog. Resolution theorem-proving yields a special kind of proof, namely a refutation. And so resolution theorem-proving is more acurately described as a *refutation procedure*. The resolution method proves a formula by showing that its negation is unsatisfiable.

How do the three non-resolution styles of proof system differ?

Axiomatic systems put theoremhood at the centre of their focus. In an axiomatic system one is given a small set of axioms, wffs which are presented by fiat as primitive theorems. In addition, an axiomatic system provides one or more, usually a minimal number, primitive rules of inference.

The rule *modus ponens* asserts that from a theorem, now taken to be a provable wff, of the form $A \Rightarrow B$, and a theorem of the form A, one

[1] *Principia Mathematica,* by Bertrand Russell and Alfred North Whitehead, Cambridge University Press, 1910-13.

[2] See *Principles of Mathematical Logic,* by David Hilbert and Wilhelm Ackermann, 1950, which is a translation of their *Grundzuge der theoretischen Logik* of 1928.

[3] Gentzen's original papers can be found in English translation in *The Collected Papers of Gerhard Gentzen,* ed ME Szabo, North-Holland, 1969.

can infer a theorem of the form B.

One writes this as

$$\frac{\mathbf{A} \qquad\qquad \mathbf{A} \Rightarrow \mathbf{B}}{\mathbf{B}}\ \textbf{(modus ponens)}$$

Modus ponens is often the sole primitive, underived rule of inference that an axiomatic system possesses. Such a rule is in fact a schema, enabling us to make an infinite number of separate inferences which conform to this pattern. The axioms are similarly schematic. We might be told that every formula of the form

$$(\mathbf{A} \Rightarrow (\mathbf{B} \Rightarrow \mathbf{C})) \Rightarrow ((\mathbf{A} \Rightarrow \mathbf{B}) \Rightarrow (\mathbf{A} \Rightarrow \mathbf{C}))$$

is an axiom, for any formulae **A**, **B** and **C**.

Since modus ponens usually plays this special role within axiomatic systems, the axioms tend to be dominated by '⇒', and the other binary connectives, like '&' and '∨', are usually defined in terms of '¬' and '⇒'.

In an axiomatic system, a proof of a wff (which is therefore a theorem) is a sequence of wffs each of which is either an axiom, or follows by modus ponens from two previous wffs in the sequence. The theorem proved is the last wff in the sequence.

A proof of a sequent, on the other hand, is a sequence of wffs each of which is an axiom, or follows from two previous wffs in the sequence by modus, or is one of the *premises* of the sequent. The proved sequent is the pair consisting of the set of premises and the last wff in the sequence.

All this is, in a sense, very unnatural, at least looked at from the point of view of everyday, unformalized reasoning. Axiomatic systems make logic look like a mathematical theory rather than like something humans use. And axiomatic systems can seems to place undue emphasis on provable wffs, and too little on provable arguments.

In ordinary reasoning we regard the connectives, as begin more or less independent from one another, and as having associated rules of inference, as embodying ways of putting propositions together and taking them apart.

For example, we think that '&' licenses us directly to

infer 'A' from 'A & B'

and to

infer 'B' from 'A & B'

and to

infer 'A & B' from 'A' togther with 'B'

This philosophical view of logic is embodied in natural deduction systems. Natural deduction systems eliminate axioms, and compensate for this loss by incorporating many more rules of inference. Ideally, a natural deduction system has one rule for introducing each connective and one rule for eliminating it. In the case of the connective '&', the first two inferenc es above are instances of the '&-Elimination' rule, and the third is an instance of the '&-Introduction' rule.

A natural deduction system views implication as a relation between a set of wffs and a single wff, much as one is inclined to in everyday logical life. An implication is represented by a sequent, a valid implication by a valid (or provable) sequent. Notice that we are now thinking of sequents, and not wffs, as being the basic objects we regard as being provable. But it is better to say that sequents, pairs of premises and a conclusion, are the objects which are proved.[4]

Taking this second view, that natural deduction systems deal essentially with sequents, one can say that a provable sequent is intended to correspond to a valid argument. Rules of inference then correspond to something more abstract, namely to transformations of valid arguments. Provable wffs are easily seem to be a special case of provable sequents, namely those having an empty set of premises.

A rule of inference in a natural deduction system is essentially a pattern involving, or more exactly a relation which holds between, a set of sequents (the premises) and a single sequent (the conclusion). Note

---

[4] Here we are treating the rules of inference as if they take us from one or more wffs - the premises - to a single wff - the conclusion. In chapter 9 we set up a natural deduction system in which the rules are defined in terms of relations between sequents - pairs in which the first element is a set of wffs and in which the second element is a single wff.

A wff-based natural deduction system is, in one sense, more natural that a sequent-based one. We tend to think of ourselves as reasoning with sentences and not with whole arguments. But there is a price to pay for this sort of naturalness. First, one has to keep track of the premises on which a wff depends. Secondly, we really needs two types of rules. In the terminology of Prawitz *Natural Deduction: A Proof-Theoretical Study* (pp. 22-23), there are *proper rules* - which handle wffs - and *improper rules* - which handle sequents. vE is an example of an improper rule. In fact, any rule which *discharges* a wff is an improper rule. So, sacrificing some naturalness, we opt for a sequent-based natural deduction system. All our rules will therefore be improper.

that we have here a perhaps unfortunate overloading of the terms 'premise' and 'conclusion'. We sometimes refer to the 'conclusion' (or 'premises') in a sequent, and we also refer to a sequent as being the 'conclusion' or as being one the 'premises' of a rule of inference. For example, from the premises

$$\mathbf{A}, \mathbf{B} \vdash \mathbf{A} \text{ and } \mathbf{A}, \mathbf{B} \vdash \mathbf{B}$$

we can infer the conclusion

$$\mathbf{A}, \mathbf{B} \vdash \mathbf{A} \;\&\; \mathbf{B}$$

by the rule '&-Introduction'. But can also say that **A** and **B** are premises of the sequent **A,B** ⊢ **A**.

A proof in a natural deduction system is most easily pictured as a tree whose root is the proved sequent. But representing a proof as a tree is typographically awkward. So we compress the tree into a sequence which incorporates labelling devices so that we can tell which sequents in the sequence are the premises of each sequent. And this results in an even better approximation to real reasoning, which is necessarily linear, at least in time.

There is, however, an asymmetry between premises and conclusion in the premises of a sequent in a natural deduction system. The premises form a set which may even be empty. *The conclusion is always a single sequent.*

This fact, though it conforms to our natural picture of implication, is also responsible for one of the awkwardnesses from which natural deduction systems tend to suffer. Take the introduction and elimination rules for '&' and '∨' in the natural deduction system which we discuss in this chapter. Suppressing reference to the premises of the sequents involved for simplicity — that is, thinking of the rules of inference as relating formula rather than sequents — the '&-Elimination' and 'v-Introduction' rules are revealed as intuitively symmetrical:

$$\frac{\mathbf{A} \;\&\; \mathbf{B}}{\mathbf{A}}\,(\&\text{-E})$$

$$\frac{\mathbf{A}}{\mathbf{A} \vee \mathbf{B}}\ (\vee\text{-I})$$

These say: from 'A & B' infer 'A', and from 'A' infer 'A ∨ B'. This symmetry breaks down when we consider the '&-Introduction' and 'v-Elimination' rules. '&-Introduction' is a nice simple rule, while 'v-Elimination' is anything but

$$\frac{\mathbf{A} \quad \mathbf{B}}{\mathbf{A}\ \&\ \mathbf{B}}\ (\&\text{-I}) \qquad\qquad \frac{\mathbf{A} \vee \mathbf{B} \quad \begin{matrix}\mathbf{A} \\ \vdots \\ \mathbf{C}\end{matrix} \quad \begin{matrix}\mathbf{B} \\ \vdots \\ \mathbf{C}\end{matrix}}{\mathbf{C}}\ (\vee\text{-E})$$

'&-Elimination' says: from a proof of 'A' and a proof of 'B', infer a proof of 'A & B'. But 'v-Elimination' says: if you proofs of 'C' from 'A' and 'B' separately, and if you have a proof of 'A ∨ B', you can then infer 'C'.

One can remedy this asymmetry in the third style of proof system, sequent calculi, but only at the cost of an unnatural view of implication — that it is not a relation holding between a premise-set and a single formula, but is rather a relation which holds between a premise-set and a conclusion set. In a sequent calculus, the proved items are *generalized* sequents, objects of the form

$$\Gamma \vdash \Delta$$

where Γ and Δ are sets (or possibly lists) of wffs.

The intuitive interpretation of a provable generalized sequent $\Gamma \vdash \Delta$ is: whenever all the wffs in Γ are true, then at least one of the wffs in Δ is true. Or, equivalently for finite Γ and Δ: whenever a conjunction of all the wffs in Γ is true, any disjunction of all the wffs in Δ is true.

In a sequent calculus, each connective has rules for introducing both into the premise-set and into the conclusion set. There are, in addition, structural rules which are independent of the properties of the connectives. One such structural rule is 'thin' which says that adding wffs to either the premise-set or conclusion-set of a provable generalized sequent yields another provable generalized sequent. That is

$$\frac{\Gamma \vdash \Delta}{\Sigma, \Gamma \vdash \Delta, \Theta}$$

where the comma represents set-theoretic union.

Sequent calculi are intermediate between axiomatic and natural deduction systems, in that they possess a single axiom:

$$\alpha \vdash \alpha$$

where α is a wff. They also possess many rules of inference, are more natural than axiomatic systems and less natural than natural deduction systems.

But the style of proof procedure that is most immediately appealing to the computer scientist is based on an approach quite different from that of the axiomatic, natural deduction, or sequent calculus approaches to logic: resolution theorem-proving.

7.3 Resolution theorem-proving

Recently, automatic theorem-proving has come to be an important subject in theoretical computing science and, via Prolog, a potentially important area of software engineering. Resolution theorem-proving involves pre-processing a wff so that it, or more accurately its negation, is in *conjunctive normal form,* and then invoking a single rule of inference exhaustively until a contradiction is derived, or, otherwise, until the rule can no longer be applied. Only in the first case, when resolution leads to a contradiction, is the negation of the original formula is shown to be contradictory and hence the original formula is proved.

We should make two points. Like an axiomatic system, the resolution theorem-proving system has a single rule of inference — the resolution rule of inference, which can be regarded as a generalization of modus ponens. But resolution theorem-proving has no axioms. It gets its apparent power from the fact that it can handle wffs only if they in conjunctive normal form. Intuitively, a wff is in conjunctive normal form if it is a conjunction of disjunctions of literals, where literals are propositional variables or negations of propositional variables. So that

$$\mathbf{P} \,\&\, \neg\, \mathbf{Q} \,\&\, (\mathbf{Q} \vee \mathbf{R} \vee \neg \mathbf{S})$$

is in conjunctive normal form, but

$$(P\&Q) \vee \neg R$$

is not.

The pre-processing required to get a wff into an equivalent wff in conjunctive normal form uses the rules that are not provable by resolution itself.

Exercise 7.1

Characterize conjunctive normal form via a BNF definition.

Exercise 7.2

Implement an acceptor in Prolog for your BNF grammar of **Exercise 7.1**.

The second point is that resolution theorem-proving, though it does generate a sequence of steps, is not quite a proof system as we have characterized it. Resolution theorem-proving begins with the wff to be proved, and attempts to derive a contradiction from its negation. It is, as we said, a refutation procedure, and not strictly a proof procedure.

Summary

A *decision procedure* for some property ψ of a set of objects is an algorithm that determines whether or not any element of that set has ψ.
A *proof procedure* is a procedure for generating proofs of provable wffs or sequents.
A procedure which is both a decision procedure and a proof procedure is called a *proof/decision procedure*.
A proof system for a formal logic is *sound* for an interpretation if every provable wff or sequent is valid according to the interpretation. A proof system for a formal logic is *complete* with respect to an interpretation if every wff or sequent which is valid according to the interpretation is provable.
We investigated the properties of the four kinds of proof system for the propositional calculus commonly encountered: the axiomatic, natural

deduction, sequent calculus, and resolution theorem-proving kinds.

8 Propositional calculus: the resolution principle

Like an axiomatic system of logic, but unlike a natural deduction system or a sequent calculus, a resolution theorem-proving system has one rule of inference, the resolution rule. Unlike an axiom system it has no axioms.

The resolution method derives its power from the amount of pre-processing that it demands be applied to a formula that it can handle. Resolution theorem-proving is a refutation procedure. It demands that the formula α to be proved be negated, and it demonstrates the unsatisfiability of the negation $\neg\alpha$ of the formula it proves. The negated formula $\neg\alpha$ must be transformed into an equivalent form in conjunctive normal form.

A formula in conjunctive normal form is a conjunction of clauses, that is, a conjunction of disjunctions of literals. A literal is either an atom, or the negation of an atom.

To make such a formula easier to handle one can think of it as a set of disjunctions of literals, that is, as a set of clauses. The original formula is then true if and only if each disjunction, or clause, in the set is true.

One can take these clauses and break them up into sets as well. Each clause can be represented as a set of literals. We call such a representation of a clause, its corresponding clause-set.

Of course not every wff is in conjunctive normal form. For example $(\mathbf{P}\&\mathbf{Q})\vee\mathbf{R}$ is not in conjunctive normal form. We saw from chapter 7 that we can characterize the class of wffs in conjunctive normal form via a BNF grammar. But is there an algorithm for transforming a wff into another wff in conjunctive normal form which is logically equivalent to it?

8.1 Transforming to conjunctive normal form — 1

For each wff of the propositional calculus there is an equivalent wff in conjunctive normal form. Here is an algorithm based on the truth-tables for constructing for any given wff, an equivalent formula in conjunctive normal form. (Not surprisingly, there is also a purely syntactical algorithm for doing the same thing, as we shall see in Section 8.2.)

To *put a wff into conjunctive normal form* is to find a wff which is logically equivalent to the original but which is such that no conjunction is 'inside' a disjunction, and which is such that its only negated sub-wffs are atomic. In effect, one must first eliminate all connectives except '¬', '∨' and '&'. But in addition we can put the resulting translation into the form of a conjunction of clauses. Our first method for putting a wff into conjunctive normal form is semantical. We refer to the wff's truth-table. Here is the recipe.

First, write out the full truth-table of the wff.

Secondly, locate the rows of the table in which the formula takes the truth-value **F**.

Thirdly, for each such row (say the i-th row), form the disjunction $\mathbf{D_i}$ of all the atoms which take the truth-value **F** and the negations of all the atoms which take the truth-value **T**.

If there is no such line — that is, the original formula *is* a tautology — then take any propositional variable which appears in the formula and let the conjunctive normal form be $\mathbf{P} \vee \neg\mathbf{P}$.

Finally, if the original formula is *not* a tautology, then conjoin all these disjunctions $\mathbf{D_i}$ of literals.

For example, to put $\mathbf{P} \Leftrightarrow \mathbf{Q}$ into conjunctive normal form, we first write out its truth-table

	P	Q	P ⇔ Q
1	T	T	T
2	T	F	F
3	F	T	F
4	F	F	T

Next, we notice that $\mathbf{P} \Leftrightarrow \mathbf{Q}$ is false in lines 2 and 3, so we form the disjunction $\neg\mathbf{P} \vee \mathbf{Q}$ from line 2, and $\mathbf{P} \vee \neg\mathbf{Q}$ from line 3, leading finally to $(\neg \mathbf{P} \vee \mathbf{Q})\ \&\ (\mathbf{P} \vee \neg \mathbf{Q})$.

Why does this algorithm work? Notice that the original formula and the new formula will be logically equivalent if and only if they take the truth-value **F** in exactly the same rows. So, choosing a row of the truth-table in which the original formula takes the truth-value **F** and taking the disjunction of the opposite truth-values assigned to the atoms in that row ensures that the two formulae take the same truth-value, **F**, for that assignment to the atoms. Now conjoining all such disjunctions (one for each row in which the original formula is false) yields a formula which is false whenever the original formula is false.

Notice also that this algorithm does not necessarily yield the *simplest* equivalent formula in conjunctive normal form. Thus taking $(\mathbf{P}\&\mathbf{Q})\vee\mathbf{R}$, the recipe yields

$$(\neg \mathbf{P} \vee \mathbf{Q} \vee \mathbf{R})\ \&\ (\mathbf{P} \vee \neg \mathbf{Q} \vee \mathbf{R})\ \&\ (\mathbf{P} \vee \mathbf{Q} \vee \mathbf{R})$$

but a simpler equivalent formula in conjunctive normal form is

$$(\mathbf{P} \vee \mathbf{R})\ \&\ (\mathbf{Q} \vee \mathbf{R})$$

8.2 Transforming to conjunctive normal form — 2

We can transform a wff into another equivalent wff in conjunctive normal form without referring to the truth-tables, which is clearly to be preferred from the point of view of theorem-proving. We proceed in three stages, or passes.

Stage 1

First eliminate occurrences of '⇔' and '⇒'.

If a formula, or subformula, is of the form

$$A \Leftrightarrow B$$

replace it by

$$(\neg A \vee B)\ \&\ (\neg B \vee A)$$

If a formula, or subformula, is of the form

$$A \Rightarrow B$$

replace it by

$$\neg A \vee B$$

So we have the transformations

$$A \Leftrightarrow B \texttt{==>} (\neg A \vee B)\ \&\ (\neg B \vee A)$$

and

$$A \Rightarrow B \Longrightarrow \neg A \vee B$$

where '==>' means *transforms to.*

Stage 2

Next, push negation signs towards the atoms using de Morgan's Laws, and eliminate double negations. Remember that at this point, the transformed formula contains only conjunctions, disjunctions and negations.

If a formula, or subformula, is of the form

$$\neg\neg A$$

replace it by

$$A$$

If a formula, or subformula, is of the form

$$\neg(A \& B)$$

replace it by

$$\neg A \vee \neg B$$

If a formula, or subformula, is of the form

$$\neg(A \vee B)$$

replace it by

$$\neg A \& \neg B$$

So we have

$$\neg\neg A \Longrightarrow A$$

$$\neg(A \& B) \Longrightarrow \neg A \vee \neg B$$

and

$$\neg(A \vee B) \Longrightarrow \neg A \& \neg B$$

Stage 3

Finally, use the distributive law(s)

$$A \vee (B \& C) \Longrightarrow (A \vee B) \& (A \vee C)$$

and

$$(A \& B) \vee C \Longrightarrow (A \vee C) \& (B \vee C)$$

repeatedly on formulae and on all subformulae to obtain a formula in conjunctive normal form.

Stage 3 frequently involves recursion to a considerable depth, and is one of the main causes of the sluggishness of the algorithm when it is coded up directly in Prolog. Here are the clauses for Stage 1.

```
            /* conjunctive normal form: stage_1 */

stage_1(iff(L,R),  and(or(neg(L1),R1),
 or(neg(R1),L1))      ) :-
 stage_1(L,L1),
 stage_1(R,R1).

stage_1(if(L,R),or(neg(L1),R1))  :-
 stage_1(L,L1),
 stage_1(R,R1).

stage  _1(and(L,R),and(L1,R1))  :-
 stage_1(L,L1),
 stage_1(R,R1).

stage_1(or(L,R),or(L1,R1))  :-
 stage_1(L,L1),
 stage_1(R,R1).

stage_1(neg(R),neg(R1))  :-
 stage_1(R,R1).

stage_1(X,X).
```

Exercise 8.1

Complete the Prolog code for transforming a wff of the propositional calculus into conjunctive normal form.

Given the query,
|- ? *cnf("(p&q)vr",C).*
the program solution to this exercise succeeds with

C = and(or(p,r),or(q,r))

which was what we found manually.

8.3 The resolution rule of inference

The resolution rule of inference handles formulae in conjunctive normal form. It is a highly intuitive rule which, in terms of disjunctions of literals, rather than sets of clause-sets is as follows.

$$\frac{\mathbf{A} \vee \mathbf{B} \qquad \neg\mathbf{B} \vee \mathbf{C}}{\mathbf{A} \vee \mathbf{C}} \ \textbf{(resolution)}$$

In words: given two disjunctions, containing a literal occurring positively in one and negatively in the other, cancel the literal in both disjunctions and collect together (in a disjunction) all the remaining literals. Call $\mathbf{A} \vee \mathbf{B}$ and $\neg\mathbf{B} \vee \mathbf{C}$ the *premises*, and $\mathbf{A} \vee \mathbf{C}$ the *conclusion*.

From the truth-tables , it is easy to see that the resolution rule is sound, that is, whenever the disjunctions in the premises are true, so is the disjunction in the conclusion. It follows that whenever the conclusion is a contradiction, so is the conjunction of the two premises. So we have the important fact that the resolution rule of inference yields a contradiction only if the conjunction of its premises is a contradiction.

If 'A' is the 'empty' disjunction, namely the disjunction of no literals, which is always false, then the rule reduces to

$$\frac{\mathbf{B} \qquad \neg\mathbf{B} \vee \mathbf{C}}{\mathbf{C}} \ \textbf{(resolution-1)}$$

which is a version of modus ponens, since the wff $\neg\mathbf{B} \vee \mathbf{C}$ is equivalent to $\mathbf{B} \Rightarrow \mathbf{C}$.

We can also represent the original rule as

$$\frac{\neg\mathbf{A} \Rightarrow \mathbf{B} \quad \mathbf{B} \Rightarrow \mathbf{C}}{\neg\mathbf{A} \Rightarrow \mathbf{C}} \ \textbf{(resolution-2)}$$

which is one version of the transitivity of 'if..then..'.

How does one use resolution to prove a wff in the propositional calculus? Suppose for example, one has the formula

$$((\mathbf{P} \vee \neg\mathbf{Q}) \,\&\, \mathbf{Q}) \Rightarrow (\mathbf{P} \vee \mathbf{R})$$

This formula is a tautology, and we expect to be able to prove it by resolution. Recall that resolution is a *refutation* procedure, so we first negate the formula, obtaining

$$\neg\, (((\mathbf{P} \vee \neg\mathbf{Q}) \,\&\, \mathbf{Q}) \Rightarrow (\mathbf{P} \vee \mathbf{R}))$$

There is, as we saw above, an algorithm for obtaining an equivalent formula in conjunctive normal form, but for now we proceed intuitively, substituting equivalents for subformula until we obtain a formula in conjunctive normal form. Thus

$$\neg\, (((\mathbf{P} \vee \neg\mathbf{Q}) \,\&\, \mathbf{Q}) \Rightarrow (\mathbf{P} \vee \mathbf{R}))$$

$\neg\, (\neg((\mathbf{P} \vee \neg\mathbf{Q}) \,\&\, \mathbf{Q}) \vee (\mathbf{P} \vee \mathbf{R})$ using 'A $\Rightarrow$ B' ==> $\neg$A $\vee$B

$(\neg\neg((\mathbf{P} \vee \neg\mathbf{Q}) \,\&\, \mathbf{Q}) \,\&\, \neg(\mathbf{P} \vee \mathbf{R}))$ using '$\neg$(A $\vee$ B)' ==> '$\neg$A & $\neg$B'

$(\neg\neg((\mathbf{P} \vee \neg\mathbf{Q}) \,\&\, \mathbf{Q}) \,\&\, \neg\mathbf{P} \,\&\, \neg\mathbf{R})$ using '$\neg$(A $\vee$ B) ==> '$\neg$A & $\neg$B'

$(\mathbf{P} \vee \neg\mathbf{Q}) \,\&\, \mathbf{Q} \,\&\, \neg\mathbf{P} \,\&\, \neg\mathbf{R}$ using '$\neg\neg$A' ==> 'A'

which is now in conjunctive normal form. We have a formula which is a conjunction of four clauses $\mathbf{P} \vee \neg\mathbf{Q}$, $\mathbf{Q}$, $\neg\mathbf{P}$ and $\neg\mathbf{R}$. The resolution rule tells us that we can resolve the first two to obtain **P** since the $\neg\mathbf{Q}$ and **Q** cancel. We now have the additional clause **P**, and the original clause $\neg\mathbf{P}$. These resolve to the *empty clause*. Since the empty clause cannot be true, the negation of the formula we began with cannot be true, which shows that the original formula cannot be false.

If this talk of empty clauses seems unsatisfactory, then resolution will appear in a clearer light if re-formulated as a rule concerning *sets of clause-sets*. As a set of clause-sets, the formula

$$(\mathbf{P} \vee \neg\mathbf{Q}) \,\&\, \mathbf{Q} \,\&\, \neg\mathbf{P} \,\&\, \neg\mathbf{R}$$

becomes

$$\{ \{\mathbf{P},\neg\mathbf{Q}\}, \{\mathbf{Q}\}, \{\neg\mathbf{P}\}, \{\neg\mathbf{R}\} \}$$

Resolving the first two clause-sets, and writing the resolvent at the beginning of the string representing the set, we obtain

$$\{ \{\mathbf{P}\}, \{\mathbf{P},\neg\mathbf{Q}\}, \{\mathbf{Q}\}, \{\neg\mathbf{P}\}, \{\neg\mathbf{R}\} \}$$

Again resolving the third and fifth clause-sets we have

$$\{\{\}, \{\mathbf{P}\}, \{\mathbf{P}, \neg\mathbf{Q}\}, \{\mathbf{Q}\}, \{\neg\mathbf{P}\}, \{\neg\mathbf{R}\}, \}$$

Therefore **(P ∨ ¬Q) & Q & ¬P & ¬R** is equivalent to a conjunction of the empty clause, the clause-set {¬**P**}, and the original clauses. But the empty clause is always false, and so therefore is the clause-set

$$\{ \{\mathbf{P}, \neg\mathbf{Q}\}, \{\mathbf{Q}\}, \{\neg\mathbf{P}\}, \{\neg\mathbf{R}\} \}.$$

Formulating of the resolution rule of inference in terms of sets of clause-sets allows us to state the rule elegantly. Given clause-sets C_k, and an atom A, the resolution rule is as follows:

$$\frac{\{ C_1, \ldots, C_i \cup \{A\}, \ldots, C_j \cup \{\neg A\}, \ldots C_n\}}{\{ C_1, \ldots, C_i \cup \{A\}, \ldots, C_j \cup \{\neg A\}, \ldots C_n, C_i \cup C_j\}}$$

So, to prove a wff in the propositional calculus by the resolution method, our strategy is therefore this.

First, negate the wff.

Secondly, turn the negated wff into an equivalent formula in conjunctive normal form.

Thirdly, represent this formula as a set of clause-sets.

Finally, use resolution exhaustively; that is. until you run out of matching literals of opposite sign or until your set of clause-sets contains the empty set, whichever is first.

If you terminate on the first condition, the original formula is not provable. If you terminate on the second, it is provable.

Clearly, this algorithm will always terminate, since each clause-set and the set of clause-sets is finite, and since an application of the resolution rule adds a new clause from which a pair of complementary literals has been removed.

8.4 Wffs in conjunctive normal form

Here are some of the tautologies of chapter 6, together with their transformations into conjunctive normal form. The wffs in conjunctive normal form were generated by our conjunctive normal form transforming program — the solution to Exercise 8.1.

Notice that each wff in conjunctive normal form is a conjunction of clauses, and that each clause in the conjunction contains a pair of complementary literals. If this were not the case, then some admissible valuation would make at least one of the clauses false.

So here is another way of testing whether a wff is a tautology: transform the wff into conjunctive normal form and check that each clause has a pair of complementary literals.

WFF	*WFF in conjunctive normal form*
(1) **p∨¬p**	p v ¬p
(2) **p⇒p**	¬p v p
(3) **¬(p&¬p)**	p v ¬p
(4) **¬¬(p)⇔p**	(¬p v p) & (¬p v p)
(5) **p&(q∨r)⇔((p&q)∨(p&r))**	
	(¬p v ¬q v p v p) & (¬p v ¬r v p v p) & (¬p v ¬q v q v p) & (¬p v ¬r v q v p) & (¬p v ¬q v p v r) & (¬p v ¬r v p v r) & (¬p v ¬q v q v r) & (¬p v ¬r v q v r) & (¬p v ¬q v p) & (¬p v ¬r v p) & (¬p v ¬q v q v r) & (¬p v ¬r v q v r)
(6) **(p⇒q)⇔(¬p∨q)**	
	(p v ¬p v q) & (¬q v ¬p v q) & (p v ¬p v q) & (¬q v ¬p v q)

Exercise 8.2

Write Prolog code for writing a wff of the propositional calculus in conjunctive normal form.

8.5 A resolution theorem-prover

A resolution theorem-prover can be made to work as follows.

Given the *negation* of the wff to be proved, in the form of a list of clauses

(1) the theorem-prover resolves the head of the list with a clause from the tail; or,

(2) if the head will not resolve with a clause from the tail, it ignores it, and begins resolving the tail of the list;

(3) if there is no resolvent then it terminates with failure, or, if the resolvent is the empty clause it terminates with success;

(4) otherwise, it places the resolvent at the head of the original set of clauses, and proceeds as at (1).

To implement the algorithm we must first transform a wff into a set of clauses. We need to transform a wff **p & (r** ∨ ¬**s)** (say) first into the form **and(p,or(r,neg(s))** then into the form [**[p]**, **[r,neg(s)]]**.

Exercise 8.3

Write clauses for **make_clause_set**/2 and **make_clause**/2. **make_clause**/2 takes a disjunction of the form **or(r,neg(s))** and represents it as a list of literals without repetitions. **make_clause_set**/2 takes a conjunction and represents it as a list of clauses.

The following clauses implement the top-level of a resolution theorem-prover.

```
resolution(Wff) :-

   cnf(Wff,CNF),
   make_clause_set(CNF,Set),
   !,
   prove(Set).

prove(CS) :-

   write(CS),nl,
   (     occurs_in([],CS)
         ;
         resolve_initial(CS,R),
         !,
         prove_1([R|CS])     ).

prove_1([H|T]) :-

   write([H|T]),nl,nl,
   occurs_in([],[H|T])
   ;
   resolve_clause(H,T,R),
   !,
   prove_1([R,H|T]).
```

Exercise 8.4

Write clauses for **resolve_initial**/2 and **resolve_clause**/2.
resolve_initial(CS,R) resolves the initial list of clauses **CS** to produce the resolvent **R**.
resolve_clause(H,T,R) resolves the head of the list of clauses with its tail to find the resolvent **R**.

Here is a photo-log of a session, using the resolution theorem-prover stored in the file *resolver,* which traces the process of resolution theorem-proving for (the negations of) some theorems of the propositional calculus. Notice that the empty clause [] is at the head of the clause-set on termination.

```
C-Prolog version 1.5
| ?- [resolver].
pc_wffs.3 consulted 2668 bytes 1.802063e-02 sec.
cnf.2 consulted 5300 bytes 3.700638e-02 sec.
clause_sets consulted 6436 bytes 5.002594e-02 sec.
resolve consulted 8692 bytes 6.700897e-02 sec.

yes
| ?- resolution("-(p->(pvr))").
[[p],[neg(p)],[neg(r)]]

[[],[p],[neg(p)],[neg(r)]]

yes

| ?- resolution("-(((pv-q)&q)->(pvr))").
[[p,neg(q)],[q],[neg(p)],[neg(r)]]

[[p],[p,neg(q)],[q],[neg(p)],[neg(r)]]

[[],[p],[p,neg(q)],[q],[neg(p)],[neg(r)]]

yes

| ?- resolution("-((p&q->r)->(p->(q->r)))").
[[neg(p),neg(q),r],[p],[q],[neg(r)]]

[[neg(q),r],[neg(p),neg(q),r],[p],[q],[neg(r)]]

[[r],[neg(q),r],[neg(p),neg(q),r],[p],[q],[neg(r)]]

[[],[r],[neg(q),r],[neg(p),neg(q),r],[p],[q],[neg(r)]]

yes
```

```
| ?- resolution("-(pvq<->-(-p&-q))").
[[p,q],[neg(p),p,q],[neg(q),p,q],[p,q,neg(p)],[neg(p)],[neg(q),
neg(p)],[p,q,neg(q)],[neg(p),neg(q)],[neg(q)]]

[[q],[p,q],[neg(p),p,q],[neg(q),p,q],[p,q,neg(p)],[neg(p)],[neg
(q),neg(p),[p,q,neg(q)],[neg(p),neg(q)],[neg(q)]]

[[],[q],[p,q],[neg(p),p,q],[neg(q),p,q],[p,q,neg(p)],[neg(p)],[n
eg(q),neg(p)],[p,q,neg(q)],[neg(p),neg(q)],[neg(q)]]

yes

| ?- resolution("-(p&q<->-(-pv-q))").
[[p],[q,p],[neg(p),neg(q),p],[p,q],[q],[neg(p),neg(q),q],[p,neg
(p),neg(q)],[q,neg(p),neg(q)],[neg(p),neg(q)]]

[[neg(q),p],[p],[q,p],[neg(p),neg(q),p],[p,q],[q],[neg(p),neg(q
),q],[p,neg(p),neg(q)],[q,neg(p),neg(q)],[neg(p),neg(q)]]

[[neg(q),p],[neg(q),p],[p],[q,p],[neg(p),neg(q),p],[p,q],[q],[ne
g(p),neg(q),q],[p,neg(p),neg(q)],[q,neg(p),neg(q)],[neg(p),n
eg(q)]]

[[neg(q)],[neg(q),p],[neg(q),p],[p],[q,p],[neg(p),neg(q),p],[p,
q],[q],[neg(p),neg(q),q],[p,neg(p),neg(q)],[q,neg(p),neg(q)],[
neg(p),neg(q)]]

[[],[neg(q)],[neg(q),p],[neg(q),p],[p],[q,p],[neg(p),neg(q),p],[
p,q],[q],[neg(p),neg(q),q],[p,neg(p),neg(q)],[q,neg(p),neg(q)
],[neg(p),neg(q)]]

yes
| ?- halt.
[ Prolog execution halted ]
```

Summary

In this chapter we discussed two methods of transforming a wff of the propositional calculus into conjunctive normal form. The first was semantical in that it relied on the truth-tables. The second was purely syntactical.

The resolution rule of inference enables one to prove a formula whose negation is in conjunctive normal form. The rule, when iterated, derives from a set of clause-sets representing the negation of the wff to be proved, another set of clause-sets containing the empty clause. Since the empty clause is never true, a set of clause-sets containing the empty clause is unsatisfiable and so the original formula will be valid.

9 Propositional calculus: natural deduction

The resolution method is a top-down refutation procedure. One begins with the formula one wants proved and works away from it, first by negating the formula, and secondly by transforming the negated formula into an equivalent formula in conjunctive normal form. Finally one tries to derive a contradiction by iteratively using a single rule of inference, the resolution rule of inference.

In axiomatic, natural deduction and sequent calculus systems of logic by contrast one works towards the formula to be proved from a given basis of axioms and/or rules of inference. So these proof systems can be thought of as bottom-up systems. In this chapter we concentrate on using a natural deduction system to generate proofs of valid sequents.

9.1 The system

In a typical natural deduction system one can introduce a formula into a proof whenever one wants. One can infer a formula from a single formula, or from several formulae, directly, as long as one keeps track on the formulae on which the inferred formula depends.

For example, from the wff **A & B** we can infer **A** on its own, and also **B** own. The truth of **A** in the first case, and of **B** in the second, depends upon the truth of whatever premises entailed **A & B**.

Similariy, if we have already inferred **A** and **B** separately, we can infer **A & B**. The new inference will have as the premises on which it depends, the union of the premises on which **A** and **B** separately depend.

Natural deduction systems capture our intuitions about such inferences. They codify our intuitions about the logical connectives, because each of the connectives has associated rules of inference which capture some of its fundamental properties.

Natural deduction systems seem to dispense with axioms — and so they are quite unlike axiomatic systems of logic — but they do so only at the cost of expanding the rules of inference of the system. In an ideal natural deduction system one has two rules for each of the logical connectives, one rule for introducing the connective and one for eliminating it from a wff.

But it is misleading to think of a natural deduction system as concerning only the logical relations between *formulae*. Formulae are implied by sets of formulae, or sets of premises.

Thus the &-Elimination rule, which we sketched above, is best thought of as asserting that from a sequent

$$\Gamma \vdash A \,\&\, B$$

we can infer the sequent

$$\Gamma \vdash A$$

and also the sequent

$$\Gamma \vdash B$$

where Γ is some fixed set of formulae.

So we have the two rules of inference, represented by the patterns:

$$\frac{\Gamma \vdash A \,\&\, B}{\Gamma \vdash A} \qquad\qquad \frac{\Gamma \vdash A \,\&\, B}{\Gamma \vdash B}$$

Other rules of inference are more complex than this, allowing one to infer a sequent not from merely one, but sometimes from two, or from even three, other sequents.

For example, &-Introduction allows one to infer $\Gamma \vdash A \,\&\, B$ from the two sequents $\Gamma \vdash A$ and $\Gamma \vdash B$. We can represent &-Introduction by the pattern

$$\frac{\Gamma \vdash A \qquad\qquad \Gamma \vdash B}{\Gamma \vdash A \,\&\, B}$$

In fact, &-Introduction licenses a slightly more general inference than this, namely that from the sequents

$$\Gamma_1 \vdash A \text{ and } \Gamma_2 \vdash B$$

one can infer the sequent

$$\Gamma_1, \Gamma_2 \vdash A \,\&\, B$$

Or, represented as before[1]

$$\frac{\Gamma_1 \vdash A \qquad\qquad \Gamma_2 \vdash B}{\Gamma_1, \Gamma_2 \vdash A \,\&\, B}$$

We speak ambiguously of the *premises* in a sequent $\Gamma \vdash A$. These will either be the wffs in Γ, or ambiguously the set Γ itself. But in a rule like &-Introduction as written above, the sequents $\Gamma_1 \vdash A$ and $\Gamma_2 \vdash B$ may themselves be referred to as *premises*, this time the premises of the *rule* for &-Introduction. Here is another ambiguity. The premises in a rule may be real sequents, whose premises and conclusion are wffs of the object language. Or they may patterns, as in our representation of the rules.[2]

Since some the rules will have at least two sequents as premises, a proof in a natural deduction system is a tree-like object. The proved sequent is the root of the *proof tree*. The internal nodes of the tree are other proved sequents, a fact which raises the question of how we get started. Which sequents are allowed at the leaves of the tree?

We need a rule which allows us simply to assert a sequent or, in other words, a rule which allows to infer a sequent from no sequents at all. So we include a rule which licenses the inference from nothing of any sequent of the form

$$A \vdash A$$

We call sequents of this form *axiom sequents*. The fact that we have them supplies the sense in which a natural deduction system, at least of the precise kind we develop in some detail, does not quite dispense with axioms altogether. We need a name for this rule, so we call it the *Rule of Axiom-Sequents*. The leaves of any proof tree are therefore sequents which result from applications of the Rule of Axiom-Sequents.

We want to represent proofs on the printed page. Trees, essentially partially ordered structures, make difficulties for us, since any textual representation of them quickly expands sideways across the page. So we

[1] Here Γ_1 and Γ_2 need not be the same set of formulae. The expression 'Γ_1, Γ_2' is, by the way, shorthand for the union of the two sets Γ_1 and Γ_2.

[2] A natural deduction system which deals with sequents rather than wffs as its basic proved objects is more manageable since in that case the dependence of a formula, or conclusion, on its premises is handled automatically.

use the device of compressing a proof into a sequence, an essentially linear structure.

In a tree representation of a proof, an application of a rule of inference at one of its internal nodes is clearly signalled by the immediate parents of the node. But in a linear compression of the tree we lose this connection between the premises of a rule and its conclusion. All we can say is that a (non-axiom) sequent must be derivable from earlier sequent(s) in the proof, though from which sequents must be left obscure, a fact which makes proofs less easier to follow. So we employ the further device of labelling members (or lines with line-numbers, because we write proofs, down the page) of the sequence, and of justifying each sequent in the proof by stating the rule applied, and citing the sequents from which it is inferred by naming the corresponding line-numbers.

What does a proof look like? One answer is: a sequence of pairs, the first element of each pair being a sequent, the second its justification. But we can break up a sequent $\Gamma \vdash A$ into its premises Γ, and conclusion A.

So another answer is this. A proof is a sequence of triples, the first element of the triple being some representation of the premises of each sequent in the proof — these are wffs in each Γ, the second being the conclusion of the each sequent A, and third its justification. With one slight modification, this is what take a natural deduction proof to be.

How do we represent an application of the Rule of Axiom-Sequents? After all, an axiom sequent is dependent on no sequents at all. Our solution to this question results in the slight modification we mentioned above. We signal axiom sequents by the string **Prem** which, as I suggest, is a signal and not a justification. Therefore we adopt the ploy of placing it where the premises of the sequent are usually represented. When we write, at some line number i

Prem **(i)** **A**

we are signalling the fact that the axiom sequent $A \vdash A$ is being introduced, or 'derived from nothing', at line (i). As we see later, the wff **A** in the application of the Rule of Assumption is named by 'i' the line at which it is introduced.[3]

[3] In all these conventions we follow the approach and the system due to Bill Newton-Smith and expounded in his book *Logic: An Introductory Course,* WH Newton-Smith,

Here, for example, is a proof, the simplest in the book, of the sequent

$$A \vdash A$$

Prem (1) A

9.2 The rules of inference

Now we lay out the ten rules of inference of the system dealing with the five connectives. We begin with the '&' rules, and continue through '¬', '⇒', '∨' and '⇔'.

&-Elimination

Both the rules for '&', &-Elimination and &-Introduction are pretty trivial. &-Elimination tells us that from a sequent of the form

$$\Gamma \vdash A \& B$$

we can infer both

$$\Gamma \vdash A$$

and

$$\Gamma \vdash B$$

So the following is a proof of

$$P \& Q \vdash P$$

which can be read as *P and Q yields P*, or as *P and Q, therefore P*.

Prem	**(1)**	**P&Q**	
1	**(2)**	**P**	**1&E**

and similarly, here is a proof of

$$P \& Q \vdash Q$$

Prem	**(1)**	**P&Q**	
1	**(2)**	**Q**	**1&E**

The first column,'**1**', in line (2) refers to the premises of the sequent which is there inferred. The '1' refers to the *conclusion* **P&Q** of the axiom sequent introduced in line (1). In the case of both of these proofs we say that line (2) is achieved by *an application of &-Elimination.* So these two proof (-representations) represent the two sequences of

Routledge and Kegan Paul, 1985

sequents

(1) P&Q ⊢ P&Q
(2) P&Q ⊢ P

and

(1) P&Q ⊢ P&Q
(2) P&Q ⊢ Q

In general, a proof in our natural deduction system will consist of a sequence of lines, the first element or column of which will either be **Prem**, or a string of numerals (and commas used to separate them). The numerals essentially name wffs, the conclusions of lines which introduce axiom-sequents. Thus every number which appears in the left-most column of a proof must be the number of a line which introduces an axiom-sequent.

In the right-most column we write the justification, a string of line-numbers followed by the name of the rule of inference applied. '1&E' says 'an application of &-Elimination using the *sequent* in line (1)'. It is important to notice that in a line of a proof

i,..,j	**(k)** α	**m,..,n Rule**

— where **i,..j,**and **m,..n** are numerals, α is a wff, and **Rule** is the name of a rule of the system — the **i,..j** name *wffs* introduced by means of the Rule of Assumptions and the **m,..n** name *sequents* via the line numbers at which they occur. The number of numerals **m,..n** which must be referred to in a justification varies from rule to rule. The more complex the rule, the more line numbers must be cited in the justification.

Remember that in using &-Elimination one must respect the associativity of '&'. The following is a proof, not a very interesting one:

Prem	**(1) P&Q&R**	
1	**(2) P**	**1&E**
1	**(3) Q&R**	**1&E**

but the following attempt at deriving

P & Q & R ⊢ P & Q

fails at line (2)

Prem	(1)	**P&Q&R**	
1	(2)	**P&Q**	**1&E{error !!}**

because P & Q & R has the same expression tree as P & (Q & R)'. All the binary logical connectives are right-associative.

&-Introduction

This simple rule tells us that from the sequents

$$\Gamma_1 \vdash A$$

and

$$\Gamma_2 \vdash B$$

infer

$$\Gamma_1, \Gamma_2 \vdash A \,\&\, B$$

Notice that inferred sequent we infer has premises which are the union of the two premises from which we infer it. For example, here is a proof of

$$P,Q \vdash P\&Q$$

Prem	(1)	**P**	
Prem	(2)	**Q**	
1,2	(3)	**P&Q**	**1,2&I**

At the third line we collect together the premises of lines (1) and (3) to form the premises of line (3). The justification requires that we cite two line-numbers, the line-numbers of the sequents from which we infer line (3) — remember that lines (1) and (2) represent two sequents **P ⊢ P** and **Q ⊢ Q.**

Combining &-Elimination and &-Introduction, we can prove

$$P\&Q\&R \vdash (Q\&P)\&R$$

Prem	(1)	**P&Q&R**	
1	(2)	**P**	**1&E**
1	(3)	**Q&R**	**1&E**
1	(4)	**(Q&R)&P**	**2,3&I**

The rules governing negation are more complex and interesting.

¬-Elimination

¬-Elimination implements one half of the double-negation tautology. It is a very simple rule to operate. Thus ¬-Elimination tells us that from

$$\Gamma \vdash \neg\neg A$$

we can infer

$$\Gamma \vdash A$$

Here is the very simplest application of ¬-Elimination, the proof of

$$\neg\neg Q \vdash Q$$

Prem	(1)	**¬¬Q**	
1	(2)	**Q**	**1¬E**

Iterating the use of the rule we can derive ¬¬¬¬P ⊢ P&¬¬P, and so

Prem	(1)	**¬ ¬ ¬ ¬P**	
1	(2)	**¬ ¬P**	**1¬E**
1	(3)	**P**	**2¬E**
1	(4)	**P&¬¬P**	**2,3&I**

¬ -Introduction

¬-Elimination codifies our intuition that double negations can always be eliminated. It is a very straightforward rule.

By contrast, ¬-Introduction is a very subtle rule. The idea is this. Suppose that we can derive a sequent of the form

$$\Gamma \vdash B \& \neg B$$

where 'B' is any wff. Since a contradiction such as 'B & ¬B' can never be true, it follows that not all the premises Γ of the sequent can be true.

Thus if we take out one of the premises quite arbitrarily, say 'P', for some 'P' ∈ Γ, then if all the remaining premises are true,[4] it must be

[4] It may be that the remaining premises can *never* be true simultaneously, but in that case '¬P' certainly follows from them taken together, since any proposition would. Any proposition follows from an an inconsistent set of premises, and indeed from any inconsistency, as we show later. Recall that a set of premises implies a proposition if and

that 'P' is false. In other words if

$$\Gamma \vdash B \& \neg B$$

is valid, so is

$$\Gamma - \{P\} \vdash \neg P$$

So ¬ -Introduction says that from a sequent of the form

$$\Gamma \vdash B \& \neg B$$

for any 'B', we can infer the sequent

$$\Gamma - \{A\} \vdash \neg A$$

for any $A \in \Gamma$.

It is a (some say counter-intuitive) truth of classical logic that from a contradiction all propositions follow.[5] Here is a proof of

$$P \& \neg P \vdash Q.$$

Prem	(1)	**P&¬P**	
Prem	(2)	**¬Q**	
1,2	(3)	**(P&¬P)&¬Q**	**1,2&I**
1,2	(4)	**P&¬P**	**3&E**
1	(5)	**¬ ¬Q**	**2,4¬I**
1	(6)	**Q**	**5¬E**

Notice that in applying ¬ -Introduction at line (5), we cite lines 2 and 4.

'**2**' names the sequent appearing at line (2) which introduces formula **¬Q.**

¬Q is removed from the premises of line (4) and then negated to form the conclusion in line (5).

We can also use ¬ -Introduction to yield a very elegant proof of the Principle of Non-Contradiction:

Prem	(1)	**P&¬P**	
	(2)	**¬(P&¬P)**	**1,1¬I**

¬ -Introduction is surprisingly useful when doing exercises in our natural deduction system, which is to say that ¬ -Introduction is a

only if whenever all the premises are true, so is the proposition.

5 This is one reason why we like our theories to be consistent. Inconsistency is so undiscriminating.

deceptively powerful rule of inference, in the sense that it is required for proving many familiar valid sequents.

⇒-*Elimination*

The rules for the conditional connective are again simple.

⇒-Introduction is also very powerful.

⇒-Elimination, however, is simply one version of the classical rule of modus ponens, which one can think of as giving the bare minimum of the properties satisfied by any conditional connective.

From

$$\Gamma_1 \vdash A \Rightarrow B$$

and

$$\Gamma_2 \vdash A$$

infer

$$\Gamma_1, \Gamma_2 \vdash B$$

Here is the simplest example of its use, in the proof of

$$P \Rightarrow Q, P \vdash Q$$

Prem	(1)	**P->Q**	
Prem	(2)	**P**	
1,2	(3)	**Q**	**1,2->E**

Even this very weak property of the conditional — captured in the rule ⇒-Elimination — is enough to enable us to derive some interesting sequents, given the other rules that we have already.

Thus from the truth of 'if P then Q' and 'if P then not Q', we can infer that 'P' must be false. That is,

$$P \Rightarrow Q, P \Rightarrow \neg Q \vdash \neg P$$

Prem	(1)	**P->Q**	
Prem	(2)	**P->¬Q**	
Prem	(3)	**P**	
1,3	(4)	**Q**	**1,3->E**
2,3	(5)	**¬Q**	**2,3->E**
1,2,3	(6)	**Q&¬Q**	**4,5&I**
1,2	(7)	**¬P**	**3,6¬I**

⇒-Introduction

⇒-Introduction is a much more subtle rule than ⇒Elimination. The idea is this: suppose that we have inferred a sequent

$$\Gamma \vdash B$$

whose premises Γ are *not empty*. Let $\Gamma = \Gamma' \cup \{A\}$, so $A \in \Gamma$. Then from

$$\Gamma \vdash B$$

we can infer that

$$\Gamma' \vdash A \Rightarrow B$$

We can justify ⇒-Introduction as follows. Whenever all the wffs in Γ' are true, then if A were true, all the wffs in Γ would be true, and therefore 'B' would also be true. If all the wffs in Γ' on its own were true, then 'if A then B' would follow.

In going from the first sequent to the second, we say that we *discharge* the wff 'A'.

The simplest application of ⇒-Introduction gives a proof of

$$\vdash P \Rightarrow P$$

Prem	(1)	**P**	
	(2)	**P->P**	**1,1->I**

From ⇒-Elimination and ⇒-Introduction together we can infer the transitivity of '⇒', that is that

$$P \Rightarrow Q, Q \Rightarrow R \vdash P \Rightarrow R$$

as follows

Prem	(1)	**P -> Q**

Prem	(2)	Q -> R	
Prem	(3)	P	
1,3	(4)	Q	1,3->E
1,2,3	(5)	R	2,4->E
1,2	(6)	P -> R	3,4->I

Notice that at line (6) where we use ⇒-Introduction, we cite the line at which the antecedent 'P' of 'P ⇒ R' is *introduced as a premise* via the sequent P ⊢ P and then the line at which the conclusion 'R' is inferred. Notice also that at line (6) we discharge the premise 'P' from the premises named in line (5).

We can now prove two important logical laws, those of Exportation and Importation respectively:

Prem	(1)	P&Q->R	
Prem	(2)	P	
Prem	(3)	Q	
2,3	(4)	P&Q	2,3&I
1,2,3	(5)	R	1,4->E
1,2	(6)	Q->R	3,5->I
1	(7)	P->(Q->R)	2,6->I
	(8)	(P&Q->R)->(P->(Q->R))	1,7->I

Prem	(1)	P->(Q->R)	
Prem	(2)	P&Q	
2	(3)	P	2&E
2	(4)	Q	2&E
1,2	(5)	Q->R	1,3->E
1,2	(6)	R	5,4->E
1	(7)	P&Q->R	2,6->I
	(8)	(P->(Q->R))->(P&Q->R)	1,7->I

As another example, we prove the result

$$\vdash (P \Rightarrow Q\&R) \Rightarrow ((P \Rightarrow Q) \& (P \Rightarrow R))$$

Prem	(1)	P->Q&R	
Prem	(2)	P	
1,2	(3)	Q&R	1,2->E
1,2	(4)	Q	3&E
1	(5)	P->Q	2,4->I
1,2	(6)	R	3&E
1	(7)	P->R	2,6->I
1	(8)	(P->Q)&(P->R)	5,7&I
	(9)	(P->Q&R)->((P->Q)&(P->R))	1,8->I

Now that we have ⇒-Introduction we have a standard strategy for proving sequents which have conditional conclusions. Just take the premises of the sequent to be proved, plus the antecedent of the conditional and prove the consequent of the conditional. Then use ⇒-Introduction to discharge the antecedent.

⇒-Introduction is both subtle and powerful. In fact, we can now prove the three axioms of the axiomatic system of logic we mentioned above. The rule of the axiom system — modus ponens — is already a part of our system.

Here are the proofs of

$$\vdash A \Rightarrow (B \Rightarrow A)$$
$$\vdash (A \Rightarrow (B \Rightarrow C)) \Rightarrow ((A \Rightarrow B) \Rightarrow (A \Rightarrow C))$$
$$\vdash (\neg A \Rightarrow \neg B) \Rightarrow (B \Rightarrow A)$$

in our system of natural deduction:

Prem	(1)	A	
Prem	(2)	B	
1,2	(3)	A&B	1,2&I
1,2	(4)	A	3&E
1	(5)	B->A	2,4->I
	(6)	A->B->A	1,5->I

Prem	(1)	A->(B->C)
Prem	(2)	A->B
Prem	(3)	A

1,3	(4)	B->C	1,3->E
2,3	(5)	B	2,3->E
1,2,3	(6)	C	4,5->E
1,2	(7)	A->C	3,6->I
1	(8)	(A->B)->(A->C)	2,7->I
	(9)	(A->(B->C))->((A->B)->(A->C))	1,8->I

Prem	(1)	¬A->¬B	
Prem	(2)	B	
Prem	(3)	¬A	
1,3	(4)	¬B	1,3->E
1,2,3	(5)	B&¬B	2,4&I
1,2	(6)	¬ ¬A	3,5¬I
1,2	(7)	A	6¬E
1	(8)	B->A	2,7->I
	(9)	(¬A->¬B)->(B->A)	1,8->I

⇒-Introduction is a powerful rule. Too powerful perhaps,[6] for it enables us to derive the so-called *paradoxes of material implication,* that is *paradoxes of the conditional:* that if 'Q' is false, then 'if Q then R' follows, for any 'R'

$$\neg Q \vdash (Q \Rightarrow R)$$

and that from 'R' follows the truth of 'if Q then R' for any 'Q'

$$R \vdash (R \Rightarrow Q)$$

Prem	(1)	¬Q	
Prem	(2)	Q	
Prem	(3)	¬R	
1,2	(4)	Q&¬Q	1,2&I
1,2,3	(5)	(Q&¬Q)&¬R	3,4&I
1,2,3	(6)	Q&¬Q	5&E
1,2	(7)	¬ ¬R	3,6¬I
1,2	(8)	R	7¬E
1	(9)	Q->R	2,8->I

[6] The *some* being philosophers of logic who want their implication to conform to their intuitions as embodied in everyone's ordinary language.

Prem	(1)	**R**	
Prem	(2)	**Q**	
1,2	(3)	**Q&R**	**1,2&I**
1,2	(4)	**R**	**3&E**
1	(5)	**Q->R**	**2,4->I**

∨-Elimination

∨-Elimination (read *vel elimination*) is the most complex of the rules of our system. That this is so results from the asymmetry, noted in Chapter 7, which arises because we have a set of formulae on the left of the '⊢ ' and only a single formula on the right of it. Here is how it works.

Suppose we can infer a sequent

$$\Gamma_1 \vdash A \vee B \quad [*]$$

Suppose also that we can infer the following sequents

$$\Gamma_2, A \vdash C \quad [**]$$

and

$$\Gamma_3, B \vdash C \quad [***]$$

that is, that from 'A' and other premises Γ_2, we can infer 'C', and from 'B' and other premises Γ_3 we infer also infer 'C'. Then ∨-Elimination allows us to infer

$$\Gamma_1, \Gamma_2, \Gamma_3 \vdash C \, [****]$$

In other words, from these three sequents [*], [**], and [***], collecting together the premises $\Gamma_1 \cup \Gamma_2 \cup \Gamma_3$ we can infer $\Gamma_1, \Gamma_2, \Gamma_3 \vdash C$.

In the justification of a line employing ∨-Elimination one has to cite *five* line-numbers, in the following order:

(1) the line containing the first sequent [*] whose conclusion is the original disjunction,

(2) the line containing the axiom-sequent introducing the first disjunct (in our case $A \vdash A$),

(3) the line containing the sequent

$$\Gamma_1, \Gamma_2 \vdash C$$

(4) the line containing the axiom-sequent introducing the second disjunct (in our case $B \vdash B$),

(5) the line containing the sequent

$$\Gamma_1, \Gamma_3 \vdash C$$

For example, here is a derivation of

$$(P \Rightarrow Q)\&(Q \Rightarrow R), P \vee Q \vdash R$$

Prem	(1)	**(P->R)&(Q->R)**	
Prem	(2)	**PvQ**	
1	(3)	**P->R**	**1&E**
1	(4)	**Q->R**	**1&E**
Prem	(5)	**P**	
1,5	(6)	**R**	**3,5->E**
Prem	(7)	**Q**	
1,7	(8)	**R**	**4,7->E**
1,2	(9)	**R**	**2,5,6,7,8vE**

Notice the form of proofs using ∨-Elimination. When, at some point in a proof, one wants to prove something from a disjunction, one must prove it from *both* disjuncts, and then collect together all the premises from both derivations, replacing the two disjuncts among the premises by the single disjunction.

The most elegant application of ∨-Elimination is this.

Prem	(1)	**PvP**	
Prem	(2)	**P**	
1	(3)	**P**	**1,2,2,2,2vE**

Lines (3) are both proofs of 'P' from 'P'. One doesn't have to write out line (2) twice.

v-Introduction

∨-Introduction, on the other hand, is as trivial as the earlier rule &-Elimination. ∨-Introduction allows us to infer

$$\Gamma \vdash P \vee Q$$

and also

$$\Gamma \vdash Q \vee P$$

for any 'Q', from

$$\Gamma \vdash P$$

We can use ∨-Elimination and ∨-Introduction to derive P ∨ Q ⊢ Q ∨ P as follows

Prem	**(1)**	**PvQ**	
Prem	**(2)**	**P**	
2	**(3)**	**QvP**	**2vI**
Prem	**(4)**	**Q**	
4	**(5)**	**QvP**	**4vI**
1	**(6)**	**QvP**	**1,2,3,4,5vE**

As another example we use ∨-Elimination and ∨-Introduction to prove one half of the one of the distributive laws.

Pv(Q&R) ⊢ (PvQ)&(PvR)

Prem	**(1)**	**Pv(Q&R)**	
Prem	**(2)**	**P**	
2	**(3)**	**PvQ**	**2vI**
2	**(4)**	**PvR**	**2vI**
2	**(5)**	**(PvQ)&(PvR)**	**3,4&I**
Prem	**(6)**	**Q&R**	
6	**(7)**	**Q**	**6&E**
6	**(8)**	**R**	**6&E**
6	**(9)**	**PvQ**	**7vI**
6	**(10)**	**PvR**	**8vI**
6	**(11)**	**(PvQ)&(PvR)**	**9,10&I**
1	**(12)**	**(PvQ)&(PvR)**	**1,2,5,6,11vE**

With the ∨- rules we can also prove the de Morgan Law:

P∨Q ⊢ ¬(¬P&¬Q)

Prem	**(1)**	**PvQ**	
Prem	**(2)**	**¬P&¬Q**	
Prem	**(3)**	**P**	
2	**(4)**	**¬P**	**2&E**
2,3	**(5)**	**P&¬P**	**3,4&I**
3	**(6)**	**¬(¬P&¬Q)**	**2,5¬I**
Prem	**(7)**	**Q**	
2	**(8)**	**¬Q**	**2&E**
2,7	**(9)**	**Q&¬Q**	**7,8&I**
7	**(10)**	**¬(¬P&¬Q)**	**2,9¬I**
1	**(11)**	**¬(¬P&¬Q)**	**1,3,6,7,10vE**

and

$$\neg(\neg P\&\neg Q) \vdash P \vee Q$$

Prem	**(1)**	**¬(¬P&¬Q)**	
Prem	**(2)**	**¬(PvQ)**	
Prem	**(3)**	**P**	
3	**(4)**	**PvQ**	**3vI**
2,3	**(5)**	**(PvQ)&¬(PvQ)**	**2,4&I**
2	**(6)**	**¬P**	**3,5¬I**
Prem	**(7)**	**Q**	
7	**(8)**	**PvQ**	**7vI**
2,7	**(9)**	**(PvQ)&¬(PvQ)**	**2,8&I**
2	**(10)**	**¬Q**	**7,9¬I**
2	**(11)**	**¬P&¬Q**	**6,10&I**
1,2	**(12)**	**(¬P&¬Q)&¬(¬P&¬Q)**	**1,11&I**
1	**(13)**	**¬¬(PvQ)**	**2,12¬I**
1	**(14)**	**PvQ**	**13¬E**

Natural deduction systems, though 'natural', do not necessarily lead to elegant derivations. Here for example is a derivation of the resolution rule of inference.

$$P\vee Q, \neg Q\vee R \vdash P\vee R$$

Prem	**(1)**	**PvQ**	
Prem	**(2)**	**P**	
2	**(3)**	**PvR**	**2vI**
Prem	**(4)**	**Q**	

Prem	(5)	¬QvR	
Prem	(6)	¬Q	
4,6	(7)	(Q&¬Q)	**4,6&I**
Prem	(8)	¬R	
4,6,8	(9)	(Q&¬Q)&¬R	**7,8&I**
4,6,8	(10)	Q&¬Q	**9&E**
4,6	(11)	¬¬R	**8,10¬I**
4,6	(12)	R	**11¬E**
Prem	(13)	R	
4,5	(14)	R	**5,6,12,13,13v**
E			
4,5	(15)	PvR	**14vI**
1,5	(16)	PvR	**1,2,3,4,15vE**

⇔-Elimination and ⇔-Introduction

⇔-Elimination and ⇔-Introduction are really straightforward re-write rules for translating between '⇔' and '⇒'. Thus ⇔-Elimination asserts that from

$$\Gamma \vdash A \Leftrightarrow B$$

we can infer

$$\Gamma \vdash (A \Rightarrow B) \,\&\, (B \Rightarrow A)$$

Similarly ⇔-Introduction allows us to infer the first of these sequents from the second. In each case the justification requires one line to be cited. Here for example are derivations which demonstrate the reflexivity and symmetry of '⇔'.

$$\vdash P \Leftrightarrow P$$

Prem	(1)	P	
	(2)	P->P	**1,1->I**
	(3)	(P->P)&(P->P)	**2,2&I**
	(4)	P<->P	**3<->I**

$$\vdash (P \Leftrightarrow Q) \Leftrightarrow (Q \Leftrightarrow P)$$

Prem	(1)	P<->Q	
1	(2)	(P->Q)&(Q->P)	1<->E
1	(3)	P->Q	2&E
1	(4)	Q->P	2&E
1	(5)	(Q->P)&(P->Q)	3,4&I
1	(6)	Q<->P	5<->I
	(7)	(P<->Q)->(Q<->P)	1,6->I
Prem	(8)	Q<->P	
8	(9)	(Q->P)&(P->Q)	8<->E
8	(10)	Q->P	9&E
8	(11)	P->Q	9&E
8	(12)	(P->Q)&(Q->P)	10,11&I
8	(13)	P<->Q	12<->I
	(14)	(Q<->P)->(P<->Q)	8,13->I
	(15)	((P<->Q)->(Q<->P))&((Q<->P)->(P<->Q))	7,14&I
	(16)	(P<->Q)<->(Q<->P)	15<->I

9.3 Soundness and completeness

Now we have eleven rules of inference in all — the rule of axiom-sequents plus two rules for each of the five connectives. The question arises: have we set up an adequate proof system for the propositional calculus? Recall the definitions of soundness and completeness we gave in chapter 7.

A proof system for a formal logic is sound for an interpretation if every provable wff or sequent is valid according to the interpretation.

A proof system for a formal logic is complete with respect to an interpretation if every wff or sequent which is valid according to the interpretation is provable.

We briefly consider the soundness of our natural deduction system. Soundness is always the simpler property of the two to prove. The interpretation of the system is given by the truth-tables. The problem is to show that every proof licensed by the system is a proof of valid sequent.

We sketch a proof of soundness which proceeds by induction on the *length of proofs* in the system. The shortest proofs have length 1. These are all of the form

Prem (1) α

for some wff α. But such a proof consists of the single sequent $\alpha \vdash \alpha$ which is obviously valid. This establishes the *base case*.

Now we make the *induction step*. We assume that all the sequents provable in k or fewer steps (that is, proofs of length k or less) are valid, and we show that, if so, all proofs of length k + 1 are valid.

The rule applied at the k+1th step is either the Rule of Assumptions or one of the ten Introduction and Elimination rules governing the connectives. If the rule is the Rule of Assumptions then the previous argument applies and the proof is valid. If it is not the Rule of Assumptions then it must be one of the ten other rules, and so we have to consider each in turn. We consider two rules and leave the rest for an exercise.

So first, suppose that the rule applied at the k+1th step is &-Elimination. Then the proof looks like this.

.

Γ (i) α & β

.

.

Γ (k+1) α j &E

where Γ is a list of line numbers each of which lies in the range 1..k and where j is also in the range 1..k. The sequent proved at line k+1 is $\Gamma \vdash \alpha$, and the rule applied is of the form

$$\frac{\Gamma \vdash \alpha \,\&\, \beta}{\Gamma \vdash \alpha}(\&E)$$

Since $1 \leq j \leq k$, $\Gamma \vdash \alpha \,\&\, \beta$ is valid by the induction hypothesis, which means that whenever all the wff in Γ are true so is α & β. By the truth-tables it follows that whenever all the wff in Γ are true so is α. So the sequent at line k+1 is valid

Suppose secondly that the rule applied at the k+1th step is $\Rightarrow$-Introduction. Here is what the proof looks like.

		.	
		.	
Prem	(i)	α	Just
		.	
		.	
Γ,i	(j)	β	
		.	
		.	
Γ	(k+1)	$\alpha \Rightarrow \beta$	i,k $\Rightarrow$E

so the rule applied is of the form

$$\frac{\alpha \vdash \alpha \qquad \Gamma,\alpha \vdash \beta}{\Gamma \vdash \alpha \Rightarrow \beta}$$

The sequents in the premises are valid, the second by the induction hypothesis. This says that whenever all the wffs in $\Gamma \cup \alpha$ are true so is β, from which it follows by the truth-table for '$\Rightarrow$' that whenever all the wffs in Γ are true so is $\alpha \Rightarrow \beta$

Similar reasoning handles the remaining eight rules of inference. So we can prove that for any proof of length 1,...., the proved sequent is valid, which gives us the soundness of the natural deduction system.[7]

Exercise 9.1

Prove the following sequents.

(1) $\vdash$ P v ¬P
(2) ¬(¬Pv¬Q) $\vdash$ P&Q
(3) ¬(¬P&¬Q) $\vdash$ PvQ
(4) P⇒Q $\vdash$ ¬PvQ

[7] For a proof of completeness, see *Logic: An Introductory Course* by WH Newton-Smith, Routledge and Kegan Paul, 1985, p. 100ff.

(5) ¬PvQ ⊢ P⇒Q

(6) P⇔Q ⊢ (¬PvQ)&(¬QvP)

(7) (¬PvQ)&(¬QvP) ⊢ P⇔Q

(8) ⊢ (P⇒(Q⇒R))⇒(P&Q⇒R)

(9) ⊢ (P&Q⇒R)⇒(P⇒(Q⇒R))

(10) P&(QvR) ⊢ (P&Q)v(P&R)

Exercise 9.2

Are the paradoxes of material implication really paradoxes? Does our ordinary use of 'if...then...' conform to the properties of '⇒' ? If not, is this a criticism of our ordinary 'if...then...' or of '⇒' or of neither?

Exercise 9.3

Convince yourself that all the rules of inference are sound. That is, they never take you from valid sequent(s) to an invalid sequent.

Exercise 9.4

We said: *In general, a proof will consist of a sequence of lines, the first element or column of which will either be 'Prem', or a string of numerals (and commas used to separate them). The numerals essentially name wffs, the conclusions of lines which introduce axiom-sequents. Thus every number which appears in such a column must be the number of a line which introduces an axiom-sequent.* Give a BNF grammar for <premise-string> the first element in a line of a proof.

Exercise 9.5

Give a BNF grammar for <justification-string> the final element in a line of a proof:
a string of line-numbers followed by the name of the rule of inference applied.

Project

Write a proof-checker (interactive or otherwise) for our natural deduction system. (A proof-checker might read text — supposedly the text of a proof — from a file, accept it if it is proof, and reject it and issue appropriate diagnostics otherwise. An interactive proof-checker traps errors as they are input from a terminal).

Summary

In this chapter we described the ten rules of inference which, together with the rules of axiom-sequent, make up the proof system of a natural deduction version of the propositional calculus. We saw how to use the rules of inference to prove valid sequents. We sketched a proof of the soundness of our system of natural deduction.

10 Predicate calculus: syntax

The classical propositional calculus enables us to assess those arguments whose validity depends only on the properties of the logical connectives '$\Leftrightarrow$', '$\Rightarrow$', '$\vee$', '&' and '$\neg$'. This is a very small class of arguments, and one is hard-pressed to come up with convincing cases where paraphrasing into the propositional calculus is a real help. The propositional calculus cannot help us to demonstrate the validity of even an argument as simple as the classic Aristotelian syllogism

All men are sinful.
Ludwig is a man.
Therefore Ludwig is sinful.

The best the propositional calculus can do is to represent the three different propositions that the syllogism contains as P, Q, and R, and to portray the argument as

P
Q
Therefore R

which is clearly not valid.

The trouble is that the propositional calculus lacks *expressive power*. It certainly cannot express typical mathematical statements. How, for example, would one represent the form of statements like the following:

Zero is the smallest natural number.
There is a smallest natural number.
Every natural number is greater than zero.
If a natural number is not zero, there is a smaller natural number.
There is no greatest natural number.

A list is either nil or is a pair consisting of an S-Expr and a list.

If the nil list has a certain property, and if, whenever a list has the property in question so does any other list consisting of an S-Expr and that list, then all lists have that property.

To exhibit the validity of our argument about Ludwig, and the forms of our six mathematical propositions, we must reveal the *internal structure* of propositions. The first thing we must do is to extend the syntax of the propositional calculus. The new, extended, language is called the *language of the predicate calculus*.

Consider the statement *zero is the smallest natural number*. This clearly tell us something about a particular natural number, namely zero. 'Zero', or '0' as we normally write it, is a *term*. The term '0' names the fixed natural number zero. And the statement tells us something about it. It says that it has a certain property, that of *being the smallest natural number*. We can express it in the form:

P(0)

(read as 'P of 0') where '0' names, or denotes, the number zero, and 'P' names, or denotes, the property of *being the smallest natural number*.

So we can see that we need two new categories of syntactic object. We need symbols which stand for (or name or denote) properties. These are called predicate-symbols. 'P' is a predicate-symbol. We need terms to stand for particular objects, like zero. The term which stands for the number zero is '0'.

Now consider the statement *there is a smallest natural number*. Clearly the statement is true, because zero is the smallest natural number, but the statement does not say so. It merely says that *there is* a smallest natural number, without specifying it. So its form must be different from that of the statement *zero is the smallest natural number*.

To exhibit the form of this proposition we need some way of asserting the existence of some object which we leave un-named. We need the *existential quantifier*. We need to be able to say

there exists ....

in a formal notation. Similarly, to exhibit the form of the false statement *every natural number is greater than zero* we need to be able to say

for all....

in a formal notation.

We also need to be able distinguish between the following four distinct cases of assertion and denial of existence.

(1) *There is a number such that...*
(2) *Everything is a number such that...*
(3) *It is not the case that there is a number such that...*
(4) *It is not the case that everything is a number such that...*

We need a formal notation for *there exists...* and *for all....* We write *there exists ...* as $\exists$. And we write *for all..* as $\forall$. We also need *variables* to associate properties with the objects whose existence is asserted.

One way of paraphrasing *there is a smallest natural number* is to write

$$(\exists\ x)\ P(x)$$

which may be read as 'for some x, P of x'. Here 'P' denotes, as before, the property of *being the smallest natural number*. But $(\exists\ x)\ P(x)$ is in fact a rather poor paraphrase of the proposition *there is a smallest natural number*, something we notice as soon as we consider the following example.

If a natural number is not zero, there is a smaller natural number.

Smaller than and *smallest* go together. We could write 'x is smaller than y' as 'S(x,y)' and 'x is equal to y' as 'E(x,y)'. To paraphrase we need to be able to say something about all natural numbers. We need the universal quantifier $\forall$. Then we can write

$$(\forall y)(\neg E(y, 0) \Rightarrow (\exists\ x)\ S(x, y))$$

which can be read as 'for all y, if y is not equal to zero then there is an x such that x is less than y'.

Smaller than denotes a relation rather than a property, one which usually denoted by '<'. We can also introduce *infix* operators like the '<' in 'x < y'. Replacing 'S(x,y)' by 'x < y' and 'E(y, 0)' by 'x = y' in

$$(\forall y)(\neg E(y, 0) \Rightarrow (\exists\ x)\ S(x, y))$$

we have the more readable

$$(\forall y)(\neg(x = y) \Rightarrow (\exists\ x)\ x < y)$$

or even

$$(\forall y)(x \neq y \Rightarrow (\exists x)\, x < y)$$

Now we can re-write our previous examples. First, *zero is the smallest natural number*.

$$(\forall y)(y \neq 0 \Rightarrow 0 < y)$$

and secondly, *there is a smallest natural number*

$$(\exists x)(\forall y)(y \neq x \Rightarrow x < y)$$

Consider also the proposition *there is no greatest natural number*. This does not tell us that some natural number has a particular property, though it appears to at first sight, because it appears to be of the form *there is...* . Rather it tells us something about *all* natural numbers.

Remembering that we assume that our variables range over the natural numbers only, the proposition may be paraphrased as

$$(\forall y)(\exists x)\, y < x$$

or equivalently

$$\neg(\exists y)(\forall x) \neg (y < x)$$

or, rendered somewhat awkwardly back into English, 'there is no natural number y, such that every natural number x fails to be greater than y'.

An important advantage of the notation we are developing is that it brings out the similarities in our four propositions (1)-(4), similarities which may not be apparent at first sight. Let the property left unspecified by *such that...* be denoted by 'Q'. Then (1)-(4) may be paraphrased as

(1) $(\exists x)\, Q(x)$
(2) $(\forall x)\, Q(x)$
(3) $\neg(\exists x)\, Q(x)$
(4) $\neg(\forall x)\, Q(x)$

Convince yourself that (3) is equivalent to $(\forall x) \neg Q(x)$, and (4) is equivalent to $(\exists x) \neg Q(x)$. We show that they are equivalent in the next chapter.

Another advantage of our notation is that it enables us to express quite clearly distinctions which are subtle and can cause confusion. Take a proposition like

Every number is less than some number.

This might appear to say something about some particular number, namely that every number is less than it, or

$$(\exists\ x)(\forall\ y)\ x < y$$

which is clearly false. What is really intended is

$$(\forall\ x)(\exists\ y)\ x < y$$

Another advantage of the new notation is that it lets us express unwieldy propositions in a simple way. So, letting 'L' denote the property of being a list, 'S' the property of being an S-expression, and 'n' stand for the nil list, our last two examples can be paraphrased as

$$(\forall\ x)(L(x) \Rightarrow (x = n \vee (\exists\ y)(\exists\ z)(S(y)\ \&\ L(z)\ \&\ x = cons(y,z)))$$

and

$$\begin{gathered} (\ P(n) \\ \& \\ (\forall\ x)(\forall y)(\forall\ z)\ (\ L(x)\ \&\ P(x)\ \&\ S(z) \Rightarrow (y = cons(z,x) \Rightarrow P(y)\ )\ )\ ) \\ \Rightarrow \\ (\forall\ x)(L(x) \Rightarrow P(x)) \end{gathered}$$

which may look complicated, but which break up and express clearly the contents of propositions which are awkward and possibly ambiguous when put into ordinary English.

In the next section, we formalize the language of the predicate calculus. We make clear what we mean by terms, predicate-symbols, and the logical symbols augmented by the quantifiers.

Exercise 10.1

Paraphrase the following statements into the language of the predicate calculus, using constants, variables, predicate- and function-symbols of your own choice:

(a) *Some men marry their deceased wife's sisters.*
(b) *Everyone is loved by someone.*
(c) *The greater of two numbers has the larger successor.*

10.1 Formal syntax

The language[1] of the predicate calculus has the following types (or syntactic domains) of symbol

(1) constants
(2) arbitrary names
(3) variables
(4) function-symbols
(5) predicate-symbols
(6) the logical connectives '$\Leftrightarrow$', '$\Rightarrow$', '$\vee$', '&', and '$\neg$'
(7) the quantifiers '$\forall$' and '$\exists$', and
(8) the brackets ')' and '('.

No symbol can belong to more than one syntactic domain. Elements of domains (1), (2), and (3) are terms. Elements of type (4) are used to build new terms out of old terms.

Predicate-symbols, elements of type (5), have their argument places filled out with terms, and therefore function like the propositional

[1] Strictly, we should speak of *languages* of the predicate calculus. All such languages have the fixed syntactic domains (6), (7) and (8) and possess an infinite number (denumerable, naturally) of variables and arbitrary names. But languages vary in the constants, function-symbols and predicate-symbols they have. They must have at least one predicate-symbol. But they can have zero, a non-zero finite number, or a denumerable infinity of both constants and function-symbols.

variables of the propositional calculus. They are an essential component which allow us to paraphrase statements into the language of the predicate calculus.

It helps to see why we have these particular categories — or syntactic domains — of terms and predicate-symbols if we consider what elements of each domain stand for. The interpretations we give to the language of the predicate calculus demand that

(1) constants, arbitrary names and variables refer (each in its own different way) to individuals in a domain of individuals,

(2) function-symbols refer to maps whose domains are the Cartesian products of the domain of individuals in the interpretation, and whose co-domain is contained in the domain of individuals in the interpretation, and

(3) predicate-symbols refer to properties or relations, that is sets of individuals) in the collection.

An interpretation is just a domain of individuals (like the natural numbers) plus required structures — functions and relations — on the domain. As we see in the next chapter, if a formula is true in an interpretation, the interpretation is said to be a model of the formula. A formula of the language which is a logical truth is found to be true in all interpretations. Every interpretation is a model of it.

To summarize.

A constant is intended to name a fixed element of the given domain in an interpretation. Every constant names a different element in a domain.

An arbitrary name denotes a fixed element of the given domain in an interpretation, but arbitrarily like a constant of integration in calculus.

A variable, however, ranges over the whole domain of the interpretation. Just why we need this distinction between constants and arbitrary names and variables will be apparent in chapter 13.

A predicate-symbol names a set or a relation on the domain and a function-symbol names a function mapping elements (or tuples of elements) of the domain to elements of the domain.

How do we represent the various syntactic domain typographically, for the purpose of writing parsers? The choices we make for the strings of categories (1) to (4) are, of course, a matter of convenience, and nothing more. For example, lower-case letters which come early in the alphabet, say 'a'...'e', are constants. Lower-case letters which come later in the alphabet, say 'n'...'u' are used for arbitrary names, and lower-case letters which come at the end of the alphabet'w'...'z' are variables. Note that we omit 'v', the symbol we use for disjunction, since we cannot write '∨' in a font available on a computer terminal.

We take function-symbols from the lower-case alphabet, say 'f'..'h'. A function-symbol is always followed by an opening bracket, a list of terms, and a closing bracket. So there will no confusion between function-symbols and arbitrary names. Predicate-symbols are upper-case letters, say 'B'..'D', 'F'...'Z'. Since we have trouble writing '∀' and '∃' we omit 'A' and 'E' from the allowed predicate-symbols. These are reserved for the quantifiers.

Strictly speaking our new language has a *countable* number of symbols of each type, just to make sure that there are enough. So we can catenate a numeral to a symbol of each type if we wish, thus expanding our our function-symbols for example, with 'f_1', 'f_2'....

Formally, a wff of the language of the predicate calculus, in our convention, is given by the following BNF definition.

```
<wff>           ::=  <C_wff> | <C_wff> ⇔ <wff>

<C_wff>         ::=  <D_wff> | <D_wff> ⇒ <C_wff>

<D_wff>         ::=  <K_wff> | <K_wff> ∨ <D_wff>

<K_wff>         ::=  <N_or_Q_wff> | <N_or_Q_wff> & <K_wff>

<N_or_Q_wff> ::=  <factor> | ¬<N_or_Q_wff> |
                     (<quantifier> <variable> ) <N_or_Q_wff>

<factor>        ::=  <atom> | ( <wff> )

<quantifier>    ::=  ∀ | ∃

<atom> ::=           <predicate_letter> |
                     <predicate_letter> ( <term_list> )

<term_list>     ::=  <term> | <term> , <term_list>

<term>          ::=  <constant> |
                     <variable> |
                     <arbitrary_name> |
                     <function_symbol> ( <term_list> )
```

where <constant>, <variable>, <predicate_symbol> and <function_symbol> are each as above.

If one drops the quantifiers — and the options

(<quantifier> <variable>) <N_or_Q_wff>

for <N_or_Q_wff>, and

<predicate_letter> (<term_list>)

for <atom>, as well as the terms, one is back to the language of the propositional calculus, the special case in which predicate-letters are unary.

This BNF definition is again easily implemented as an program in Prolog, using the Prolog grammar rules. The parser generates an expression tree in the usual way. It is perhaps best to choose a representation in which the expression tree is not binary, so that the wffs

P,

P(x,y,f(g(x),h(a,f(x))))

and

$$(\forall x)(\forall y)P(x,y) \Leftrightarrow \neg(\exists x)(\exists y)\neg P(x,y)$$

are represented as shown in the following session.

```
| ?- wff(X,"P",[]).
X = P .
yes

| ?- wff(X,"P(x,y,f(g(x),h(a,f(x))))",[]).
X  =  P([x,y,f([g([x]),h([a,f([x])])])])  .
yes

| ?- wff(X,"(Ax)(Ay)P(x,y)<->-(Ex)(Ey)-P(x,y)",[]).
X =
iff(all(x,all(y,P([x,y]))),neg(some(x,some(y,neg(P([x,y])))))
) .
yes
```

Here is the Prolog code for the new clauses of **pd_wffs.**

```
                    /* pd_wffs */

:- consult('pd_terminals').

 conjunction_wff(Tree) -->
  quantified_or_negated_wff(Tree_Left),
  (     and,!,
        conjunction_wff(Tree_Right),
        {Tree = and(Tree_Left,Tree_Right)}
        ;
        {Tree = Tree_Left} ).

 quantified_or_negated_wff(Tree) -->
        negation_sign,
        quantified_or_negated_wff(Tree_Right),
        {Tree = neg(Tree_Right)}
        ;
        quantifier(Quantifier,Variable),
        quantified_or_negated_wff(Tree_Right),
        {Tree =.. [Quantifier | [Variable , Tree_Right ] ]}
        ;
        factor(Tree).

 factor(Tree) -->
  left_bracket,!,
  wff(Tree),
  right_bracket
  ;
  atomic_proposition(Tree).

 atomic_proposition(Tree) -->
  predicate_symbol(Symbol),
        (     left_bracket,!,
              term_list(Term_List),
              right_bracket,
              {List = [Symbol , Term_List],
               Tree =.. List}
              ;
              {Tree = Symbol}          ).
```

```
quantifier(Quantifier,Variable) -->
   left_bracket,
   (universal_quantifier(Quantifier),!
    ;
     existential_quantifier(Quantifier) ),
    !,
     variable(Variable),!,
    right_bracket.


term_list(Term_List) -->
   term(Term),
   (     comma,!,
         term_list(Term_List_1),
         {Term_List = [Term | Term_List_1]}
         ;
         {Term_List = [Term]}    ).


term(Term) -->
   functional_term(Term),!
   ;
   variable(Term),!
   ;
   arbitrary_name(Term).

functional_term(Term) -->
   function_symbol(Symbol),
         (      left_bracket,!,
                term_list(Term_List),
                right_bracket,
                {List = [Symbol, Term_List],
                Term =.. List}
   ;
   {Term = Symbol}    ).
```

and for **pd_terminals**

```
                    /* pd_terminals */

predicate_symbol(P_rep,[P|T],T) :-
  P > 64,!,                /* upper-case letters */
  not(P = 65),!,
  not(P = 69),!,           /* omit quantifiers */
  P < 91,!,
  name(P_rep,[P]).

variable(P_rep,[P|T],T) :-
  P > 116,!,               /* u,w,..,z */
  not(P = 118),!,
  P < 123,!,
  name(P_rep,[P]).

arbitrary_name(P_rep,[P|T],T) :-
  P > 108,!,               /* m,...,t */
  P < 117,!,
  name(P_rep,[P]).

function_symbol(P_rep,[P|T],T) :-
  P > 96,!,                /* a,...,l */
  P < 109,!,
  name(P_rep,[P]).

universal_quantifier(all) --> "A".
existential_quantifier(some) --> "E".

iff --> "<->".
if --> "->".
or --> "v".
and --> "&".
negation_sign --> "-".
left_bracket --> "(".
right_bracket --> ")".
comma --> ",".
```

Exercise 10.2

Let predicate- and function-symbols be strings of lower-case letters (and underscore characters) followed optionally by a numerals.
Give EBNF rules for <predicate-symbol> and <function-symbol> and re-implement the parser for **pd_wffs**.
A ground term is a term which contains no occurrences of variables. Give a BNF grammar for <ground_term>.

Summary

We explained why the language of the propositional calculus has insufficient expressive power. We extended the syntax of the propositional calculus with non-zero-ary predicate symbols, and with terms and quantifiers. We briefly discussed the process of paraphrasing from English into the formal language of the predicate calculus. We gave a context-free grammar for the language of the predicate calculus which led immediately to an implementation in Prolog of a parser for the language.

11 Predicate calculus: semantics

To decide whether or not a wff in the propositional calculus is a tautology we write out its truth-table. Each row of the truth-table represents an assignment of truth-values to the propositional variables occurring in the wff. Each row of the truth-table represents a set of possible worlds, a set of admissible valuations, a way the world, described by the propositional variables, can be.

A wff is a tautology if and only if it is true in all possible worlds. It is decidable whether or not a wff is a tautology, since the truth-table for any wff is finite.

11.1 Predicate calculus: interpretations

Is there a corresponding decision procedure for the validity of a wff in the predicate calculus? Clearly the truth-tables will not deliver. Consider the wff

$$(\forall x)(P(x) \Rightarrow P(x))$$

or, 'for all x, if P of x then P of x', which is intuitively valid. But as far as truth-tables are concerned '$(\forall x)(P(x) \Rightarrow P(x))$' is an atomic proposition, and so its truth-value may be either **T** or **F**.

To determine whether '$(\forall x)(P(x) \Rightarrow P(x))$' is valid or not we need to look *inside* it. The predicate calculus requires a more 'penetrating' semantics than does the propositional calculus.

The semantics of the propositional calculus really deals only with the logical connectives. We expect an adequate semantics for the predicate calculus to deal with each of the syntactic domains — terms, predicate-symbols, quantifiers — detailed in the last chapter.

Constants, and arbitrary names, name — or denote — fixed individuals. And so we need domains of individuals in our interpretations, as we noted in the last chapter. Variables to *range over* individuals. Therefore, we need interpretations which consist, first of all, of non-empty domains of individuals.

But given any such domain we must then provide interpretations for the predicates and function-symbols of the language of the predicate calculus. Finally, we must specify the conditions under which a wff which contains occurrences of the logical connectives, and/or the quantifiers, is true or is false in the interpretation.

By an *interpretation* I of the language of the predicate calculus, we mean a domain D, together with an individual in the domain assigned to each constant and arbitrary name, a relation assigned to each predicate-symbol, and a function to each function-symbol.

An interpretation, with its additional structure and assignment of truth-values to any wff of the language of the predicate calculus, is like an admissible valuation of the propositional calculus, it is like that which, in the propositional calculus, generates a row of the truth-table.

A wff of the propositional calculus is logically true if and only if it is true in all the admissible valuations of the language. This comes down to its being true in all the rows of its truth-table. Similarly, a wff of the predicate calculus is logically true if and only if it is true in all the much more highly structured and diverse *interpretations of the language of the predicate calculus*. Now we make these intuitive remarks more precise.

An interpretation of the predicate calculus is a *non-empty* set D together with assignments specified as follows:

(1) any constant 'c', (or arbitrary name 'm') of the language names (or denotes) a fixed element c (or m) of D,

(2) a variable 'x' of the language is allowed *to range freely over* elements of D, and together with functions on D such that

(3) a function-symbol 'f' of arity n of the language, denotes an n-ary function[1]

$$f: D^n \rightarrow D,$$

So that $f(t_1,...t_n)$ denotes some individual $t \in$ D. Or, equivalently

$$f(t_1,...t_n) = t$$

---

1 D^n means the cartesian product $D \times D \times ... \times D$, n times.

(4) a predicate-symbol 'P' of arity n of the language, denotes an n-ary relation

$$P: D^n$$

The sentence '$P(t_1,...t_n)$' takes the truth-value **T** in the interpretation if and only if

$$< t_1,...t_n > \in P.$$

(If n = 1, *P* is a property. If n = 0, P is a propositional variable. Otherwise *P* is relation.)

(5) the five logical connectives common to both the propositional and the predicate calculi take their usual truth-functional meanings,

(6) the quantifiers in the forms '$(\forall x)$' and '$(\exists x)$' are interpreted to mean *for all x in D* and *for at least one x in D* respectively.

Any given interpretation is a mapping which sends any wff without free variables — that is, any closed wff, or a wff in which all the occurrences of variables are bound by quantifiers — into {**T**,**F**}.

A wff is logically true (or valid) if and only if it takes the truth-value **T** in all (possible) interpretations. Some formulae are logical truths of the predicate calculus.

Example

Thus consider the wff

$$(\forall x)(P(x) \Rightarrow P(x))$$

Take any interpretation I with domain D. Let 'x' take the value $a \in P$, so that 'P' is true of a, as captured in our first diagram. Since 'P(a)' takes the truth-value **T**, then '$P(a) \Rightarrow P(a)$' takes the truth-value **T**.

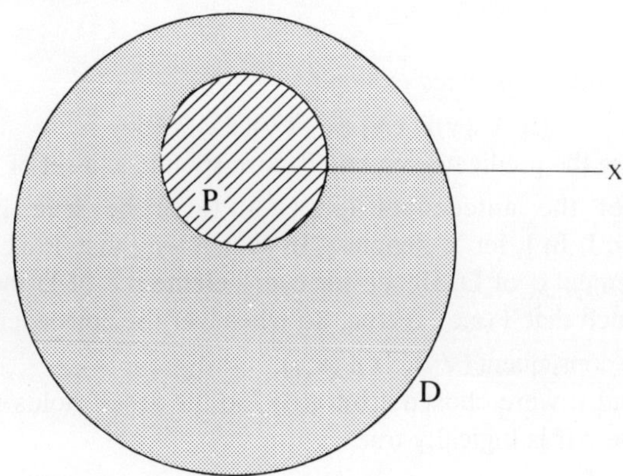

If 'P(a)' takes the truth-value **F**, which it will whenever a ∉ D as represented by the next diagram, then 'P(a) ⇒ P(a)' also takes the truth-value **T** by the truth-tables for '⇒'.

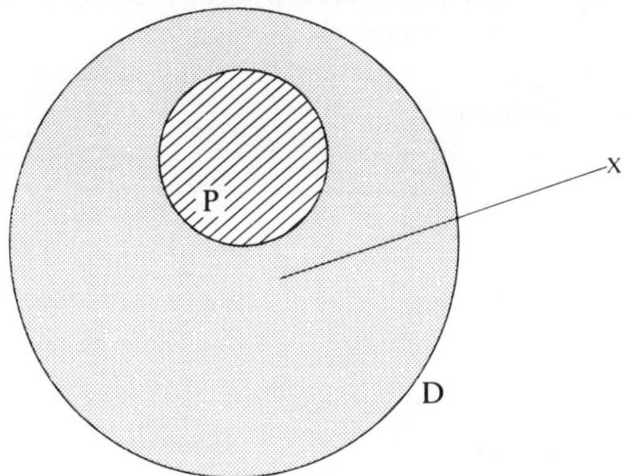

So, whichever a ∈ D we take, the wff is true in I. Our choices of D, I and 'a' and 'x' were arbitrary, so

$$(\forall x)(P(x) \Rightarrow P(x))$$

is true in all interpretations. Hence it is valid.

Example

Consider a more complicated example:

$$(\exists x)(\forall y)P(x,y) \Rightarrow (\forall y)(\exists x)P(x,y)$$

In this case the predicate-symbol 'P' corresponds to a set of ordered-pairs *P*. Let the antecedent $(\exists x)(\forall y)P(x,y)$ be true in some interpretation I. In I, let 'x' denote a, in such a way that 'P(a,c)' is true for every element c of D. Hence for every element c of D there is an element a, such that 'P(a,c)' is true. So whenever the antecedent is true in I, so is the consequent $(\forall y)(\exists x)P(x,y)$.

D, I, a, and c were chosen arbitrarily, so the result holds for *all* I. Therefore the wff is logically true.

Example

As a final example consider

$$(\forall y)(\exists x)P(x,y) \Rightarrow (\exists x)(\forall y)P(x,y)$$

If this wff is to be false in an interpretation I, the antecedent must be true in I, and the consequent false in I. Consider an interpretation whose domain D = {a, b, c}. Let the predicate-symbol 'P' denote the relation

$$P = \{ \langle a,b \rangle, \langle b,c \rangle, \langle c,a \rangle \}$$

In the following diagram we represent P's holding between a and b by an arrow directed from a to b.

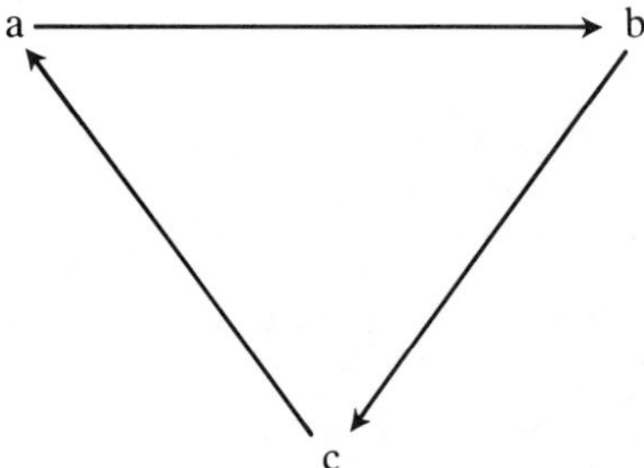

In this case, $(\forall y)(\exists x)P(x,y)$ is clearly true, but $(\exists x)(\forall y)P(x,y)$ is false so that

$$(\forall y)(\exists x)P(x,y) \Rightarrow (\exists x)(\forall y)P(x,y)$$

cannot be logically true.

Exercise 11.1

Give an interpretation whose domain consists of only the two elements a and b, which constitutes a counter-example to the wff

$$(\forall y)(\exists x)P(x,y) \Rightarrow (\exists x)(\forall y)P(x,y)$$

These examples show that we can demonstrate the validity or invalidity some predicate calculus wffs by considering possible interpretations. This suggests the following important question. Is there a decision procedure for validity in the predicate calculus, as there is for the propositional calculus? Or, to put the question in another way, if a wff is *not valid* can we generate counter-examples *automatically*?

The answers are that there isn't and we can't. One cannot write a program to run on a computer — no matter how fast the computer or how large its memory — which will tell you whether or not a wff of the predicate calculus, presented as input, is logically true.

In fact, the situation is slightly more complicated. If a wff of the predicate calculus is logically valid, then one *can* prove it automatically. A computer program can be written which grinds out proofs. But there will be at least one non-valid wff (and in fact an infinite number of such wffs) such that if the program is asked to grind out the proof of the wff, the program will fail to terminate, and so deliver no answer. Because of the first fact — that a computer program can be written which grinds out all proofs — the predicate calculus is said to be *semi-decidable*. A computer can, in principle, do half of what is required for a decision procedure, but it cannot always detect non-validity.

Nevertheless, any wff of the predicate calculus either is or is not logically true. For any wff of the predicate calculus one can either produce a counter-example as in the case of

$$(\forall y)(\exists x)P(x,y) \Rightarrow (\exists x)(\forall y)P(x,y)$$

that is, an interpretation in which it is false, or, if the wff is a logical truth, one can produce an argument of the sort that we did produce in the cases of

$$(\forall x)(P(x) \Rightarrow P(x))$$

and

$$(\forall y)(\exists x)P(x,y) \Rightarrow (\exists x)(\forall y)P(x,y)$$

Those arguments are informal proofs of the formulae concerned. The important point is that there is no algorithm for deriving counter-examples to wffs that are not valid.

In the next chapter, we begin to deal with the formalisation of the idea of proof in the predicate calculus. Unsurprisingly, it turns out that there are logical systems in the usual styles — axiomatic systems, natural deduction systems, sequent calculi — in which every logical truth is provable. But because of the semi-decidability of the predicate calculus, there are also complete *automatic theorem-provers* for the predicate calculus. (Even though there aren't complete automatic non-theorem *falsifiers.*) The one most commonly used theorem-provers in computation combines the resolution method of chapter 8, with a method of handling variables known as *unification.* Prolog embodies a restricted resolution theorem-prover of exactly this type.

11.2 Unsatisfiable wffs in the predicate calculus

It seems that to decide whether a wff of the predicate calculus is valid we must examine *all* its possible interpretations, since a wff is valid if and only if it is true in all its interpretations. But we cannot consider all the interpretations of the language.

Can we cut down the set of interpretations we need to examine in order to test a wff for validity? The answer is that we can. We can eliminate irrelevant interpretations. To do so most efficiently we must transform a given wff into a manageable form. But first some motivation.

If a wff is valid then it is true in all its interpretations. A wff is valid iff its negation is false in all interpretations. If a wff is false in all its interpretations we say that it is *unsatisfiable.* So we can demonstrate the validity of a wff by demonstrating its unsatisfiability.

Example

Consider the following very simple wff

$$P(c) \Rightarrow P(c)$$

which is not only very simple, but contains no variables.

First negate the wff and obtain

$$\neg(P(c) \Rightarrow P(c))$$

which is equivalent to

$$\neg\neg(P(c) \,\&\, \neg P(c))$$

which in turn is equivalent to

$$P(c) \,\&\, \neg P(c)$$

Thus $P(c) \Rightarrow P(c)$ is valid iff $P(c) \,\&\, \neg P(c)$ is unsatisfiable.

In any interpretation of the wff $P(c) \Rightarrow P(c)$, the constant c denotes a fixed individual. Since, in every interpretation, either $c \in P$ or $c \notin P$, but not both, it follows that $P(c) \,\&\, \neg P(c)$ is unsatisfiable and therefore that

$$P(c) \Rightarrow P(c)$$

is valid.

Of course, we hardly needed to go through this elaborate process for $P(c) \Rightarrow P(c)$ since the wff contains no variables. P(c) is, in a sense, atomic. P(c) can be treated as some Q (say) of the propositional calculus. The same applies to more complex cases like like

$$P(f(a,g(b))) \Rightarrow P(f(a,g(b))) \vee Q(g(d))$$

We need only appeal to the properties of the logical connectives, as revealed by the truth-tables of the propositional calculus, to decide the validity of a wff which contains no variables.

Example

Consider another example, this time a wff which does contain a variable:

$$(\forall x)P(x) \Rightarrow P(c)$$

Negating we get

$$\neg((\forall x)P(x) \Rightarrow P(c))$$

which is equivalent to

$$\neg\neg((\forall x)P(x) \,\&\neg\, P(c))$$

which is equivalent to

$$(\forall x)P(x) \,\&\neg\, P(c)$$

In every interpretation, either $c \in P$ or $c \notin P$, but not both, so that it follows that either $\neg$ P(c) — the former case — or $(\forall x)P(x)$ — the latter case — is false. So $(\forall x)P(x) \mathbin{\&}\neg P(c)$ is unsatisfiable and that $(\forall x)P(x) \Rightarrow P(c)$ is valid.

There is an important point that both of these simple examples exhibit. In any interpretation of the wffs each distinct constant denotes a unique element of the domain of the interpretation. This suggests that when considering the validity or the unsatisfiability of a wff, we can restrict the interpretations we consider to those that are composed of the *terms* that can be built up from terms in the formula. Instead of considering *all interpretations,* we need consider only a very restricted class whose elements are *syntactic objects* taken from the syntactic domain of terms, in fact those terms that are built up from constants and function-symbols found in the wff.

The elements of all the interpretations which can be considered to be relevant to the validity of a wff are fixed. This does not mean that there is just one relevant interpretation, because the denotations of predicates vary from one interpretation to the next. So in the cases of the last two formulae we need only consider interpretations with the fixed domain $\{c\}$. And there are only two relevant interpretations to consider. In the first, $c \in P$, and in the second $c \notin P$.

Exercise 11.2

Use the technique of this section to demonstrate the validity of

$$P(c) \Rightarrow (\exists x)P(x)$$

We know that we can prove a wff valid by showing that its negation is unsatisfiable. We know how to go about demonstrating unsatisfiability for some wffs, particularly for those that contain no variables, and those in which there is at least one constant and in which the unsatisfiable wff contains only universal quantifiers. We have seen that in these cases we need consider only certain relevant interpretations, those whose elements are built up out of constants and function-symbols appearing in the wffs concerned. Now we move on to the general case. We want to be able handle all wffs including those with exitential quantifiers and, more interestingly, we want to arrive at an *automatic* procedure for demonstrating unsatisfiability. To achieve all this we must put our wffs into a certain standard form. We proceed via a wff in *Prenex normal form*, and then via its *Skolem standard form.*

11.3 Prenex normal form

A wff is in *prenex normal form* if and only if it has the form

$$(Q_1 \text{ variable}_1)...(Q_n \text{ variable}_n)\ \psi$$

where Q_i for $1 \leq i \leq n$ is one of the two quantifiers, and ψ is a wff in which there are *no* occurrences of quantifiers. In general, ψ will contain predicate symbols, function symbols, constants, arbitrary names, variables, and logical connectives (and brackets).

ψ is said to the *matrix* of the wff.

Thus

$$(\forall x)(\exists y)(P(x) \Rightarrow Q(x,y))$$

is in prenex normal form but

$$(\forall x)(P(x) \Rightarrow (\exists y)Q(x,y))$$

is not.

Any formula can be 'put into' prenex normal form, that is any formula can be transformed into a logically equivalent wff which is in prenex normal form. To achieve this one simply moves the quantifiers outwards in such a way that all the occurrences of '¬' which appear in the final wff are made to occur in the matrix ψ.

There are many algorithms for putting a wff into prenex normal form each of which is similar to the algorithm for putting a formula of the propositional calculus into conjunctive normal form. Here is the one which is most similar to our recipe for transforming a propositional calculus wff into conjunctive normal form. In fact, it also puts the matrix into conjunctive normal form, which we will find convenient in any case. It is not the most efficient algorithm, but it is straightforward. We proceed in four stages.

Stage 1

First, eliminate occurrences of '⇔' and '⇒', in favour of '¬' and '∨', exactly as we did when we transformed a propositional calculus wff into conjunctive normal form.

Stage 2

Next push negations inwards using de Morgan's Laws and eliminate double negations as before. But, also use the additional rules

$$\neg(\forall x)P \implies (\exists x)\neg P$$

and

$$\neg(\exists x)P \implies (\forall x)\neg P$$

where '==>' means *transforms to*, as before.

Stage 3

Rename the bound variables in the wff so that all the variables bound in a quantifier are different.

For example, the wff

$$(\forall x)(\neg P(x) \vee (\exists x)Q(x))$$

should be transformed to (say)

$$(\forall x)(\neg P(x) \vee (\exists y)Q(y))$$

where the third and fourth occurrences of 'x' have been changed to occurrences of 'y'.

Stage 4

Move the quantifiers outwards. Thus

$$(\forall x)P \mathbin{\&} R \implies (\forall x)(P \mathbin{\&} R)$$
$$P \mathbin{\&} (\forall x)R \implies (\forall x)(P \mathbin{\&} R)$$
$$(\forall x)P \vee R \implies (\forall x)(P \vee R)$$
$$P \vee (\forall x)R \implies (\forall x)(P \vee R)$$

and

$$(\exists x)P \mathbin{\&} R \implies (\exists x)(P \mathbin{\&} R)$$
$$P \mathbin{\&} (\exists x)R \implies (\exists x)(P \mathbin{\&} R)$$
$$(\exists x)P \vee R \implies (\exists x)(P \vee R)$$
$$P \vee (\exists x)R \implies (\exists x)(P \vee R)$$

Re-naming the bound variables (Stage 3) is important in cases like this one since

$$(\forall x)(P(x) \mathbin{\&} (\exists x)Q(x))$$

is not equivalent to

$$(\forall x)(\exists x)(P(x) \,\&\, Q(x))$$

The result is a wff in prenex normal form. It is easy to see a wff is logically equivalent to the wff in prenex normal form[2] which is the result of applying the algorithm since each of the stages transforms a wff to another wff which is logically equivalent. (We prove the validity of some of the steps in stage 4 in the chapter 13.)

11.4 Skolem standard form

Now we go further and consider a transformation which does *not* genetate a logically equivalent wff.

This second and additional transformation does have the important property that the original wff is unsatisfiable if and only if its transformed wff — which appears in the form of a set of clauses (or clause-sets) — is also a unsatisfiable.

We put a wff into *Skolem standard form* by first putting it into prenex normal form and then by applying two further steps. A wff in prenex normal form consists of a string of quantifiers in front of a matrix, the matrix consisting of conjunctions, disjunctions and/or negations of atomic formulae. The matrix contains no quantifiers and can be treated as if it were a wff of the propositional calculus with free occurrences of variables, generally. Therefore we can put the matrix into an equivalent in conjunctive normal form.

Stage 5

So that, in obtaining an equivalent wff in Skolem standard form, we first

[2] Notice that the simple tranformation of Stage 4 does not apply to '⇒' and '⇔', which is why we need Stage 1.

$$(\forall x)P \Rightarrow R \implies (\forall x)(P \Rightarrow R)$$

isn't a valid transformation , since

$(\forall x)P \Rightarrow R$ is equivalent to $\neg(\forall x)P \vee R$

$\neg(\forall x)P \vee R$ is equivalent to $(\exists x)\neg P \vee R$

$(\exists x)\neg P \vee R$ is equivalent to $(\exists x)(\neg P \vee R)$ by Stage 4

$(\exists x)(\neg P \vee R)$ is equivalent to $(\exists x)(P \Rightarrow R)$

So the correct transformation is

$$(\forall x)P \Rightarrow R \implies (\exists x)(P \Rightarrow R)$$

put the matrix of the wff in prenex normal form into *conjunctive normal form*.

Stage 6

The final stage is the most obscure and the most interesting.

For up until now, the execution of each stage has resulted in a transformed formula which is logically equivalent to the original formula. Now we relax this restriction and transform the current formula into one which, though it is not logically equivalent to the original formula, will be unsatisfiable (and so a contradiction) if and only if the current wff is unsatisfiable. This final stage does not maintain logical equivalence, but it does maintain contradictoriness.

Stage 6 consists in replacing each universally quantified variable by a corresponding arbitrary name, and each existentially quantified variable by a *Skolem function*. When this substitution has finished, all the existential quantifiers are eliminated.

An appropriate Skolem function is an arbitrary name or an arbitrarily chosen function-symbol *which does not occur in the wff.*

It is an arbitrary name if no universal quantifiers appear to the left of the existential quantifier in the prenex normal form form of the wff.

Otherwise, it is a function of those variables corresponding to the universally quantified variables which appear to the left of the existential quantifier in the prenex normal form of the wff. Note that the substitutions of arbitrary names for the bound variables begins at the left of the wff in prenex normal form.

Example

The effect of Stage 6 on

$$(\exists x)P(x)$$

might be

$$P(a)$$

since there are *no universal quantifiers to the left* of the '(∃x)'.

Example

The effect of Stage 6 on the slightly more complex wff

$$(\forall x)(\exists y)P(x,y)$$

might be

$$(\forall x)P(x,f(x))$$

where 'f(x)' replaces 'y' because '$(\forall x)$' is to the left of '$(\exists y)$'.

Example

The effect of Stage 6 on

$$(\forall u)(\exists w)(\forall x)(\forall y)(\exists z)P(x,y,w,u,z)$$

might be

$$(\forall u)(\forall x)(\forall y)P(x,y,f(x),u,g(u,x,y))$$

since we first substitute (say) 'f(x)' for 'w', then 'g(u,x,y)' for 'z'.

Example

Consider the following example, in which we trace Stages 1 through 6. Let the original wff be

$$(\forall x)(P(x) \Rightarrow (\exists x)(Q(x) \ \& \ R(x)))$$

After Stage 1 we have

$$(\forall x)(\neg P(x) \vee (\exists x)(Q(x) \ \& \ R(x)))$$

Stage 2 does not effect the wff, but after Stage 3 we have

$$(\forall x)(\neg P(x) \vee (\exists y)(Q(y) \ \& \ R(y)))$$

Moving the quantifiers outwards — Stage 4 — yields

$$(\forall x)(\exists y)(\neg P(x) \vee (Q(y) \ \& \ R(y)))$$

Stage 5, transforming the matrix $(\neg P(x) \vee (Q(y) \ \& \ R(y)))$ into conjunctive normal form, yields

$$(\forall x)(\exists y)(\ (\neg P(x) \ \& \ Q(y)) \vee (\neg P(x) \ \& \ R(y))\ )$$

Stage 6, eliminating the quantifiers and introducing the Skolem function f(x) for 'y', yields

$$(\forall x)(\ (\neg P(x) \ \& \ Q(f(x))) \vee (\neg P(x) \ \& \ R(f(x)))\ )$$

Now we adopt a *convention*. Instead of writing the final version of this wff as

$$(\forall x)(\ (\neg P(x) \ \& \ Q(f(x))) \vee (\neg P(x) \ \& \ R(f(x)))\ )$$

we *drop the universal quantifier*, leave the 'x' free and interpret the free 'x' as if it were universally quantified. So we prefer to write

$$((\neg P(x) \,\&\, Q(f(x))) \vee (\neg P(x) \,\&\, R(f(x))))$$

as a notational convenience.

We can think of this as *a set of clause-sets*. Each clause-set is a set of atoms, which may be either propositional variables, as in the propositional calculus, or predicate-symbols complete with terms, as they generally are in the predicate calculus:

$$\{ \{\neg P(x) , Q(f(x)) \},\{\neg P(x) , R(f(x)) \} \}$$

The important point is this. *A wff is unsatisfiable (is a contradiction) iff its Skolem standard form is also.* Or equivalently, *a wff is satisfiable iff its Skolem standard form is also.*

This result is called *Skolem's Theorem*[3]. If a wff is transformed into an Skolem standard form which is then unsatisfiable, the original wff was similarly unsatisfiable.

We still have not finished. We cannot check whether or not a wff is true in every domain, since a domain may contain an infinite number of individuals. Checking the truth-value of wff by checking all domains would be like trying to write out a truth-table with an infinite number of propositional variables.

We have done nothing with the free variables, the implicit universal quantifiers. It so happens we can eliminate these by replacing occurrences of bound variables with appropriate ground terms — constants and arbitrary names and functions thereof (if necessary) — and then throwing away the universal quantifiers in front of the matrix. To do this we have to narrow down our choice of appropriate constants and arbitrary names, exactly as we did when considering our examples in Section 11.2. In fact we make our choice from the terms of the

---

[3] To see that Skolem's Theorem holds, let α be the original wff in prenex normal form and let *Skolem*(α) be the result of putting α into Skolem standard form. Since α is logically equivalent of the original wff we need only show that α and *Skolem*(α) are either both unsatisfiable or both satisfiable. Clearly if α contains no occurrences of existential quantifiers then the result follows, so suppose that α contains occurrences of existential quantifiers. Suppose first that α is unsatisfiable. If *Skolem*(α) is satisfiable we can replace each occurrence of a Skolem function with an appropriate existentially quantified variable, which implies that α is satisfiable, contradicting our assumption. Suppose on the other hand α is satisifiable. Then in an interpretation which satisifies α, let c be one of the individuals which may be substituted for the leftmost existentially quantified variable in α. With an appropriate choice of Skolem function, such that the function maps the variables it contains to c, *Skolem*(α) can be made to be satisified. Hence if α is satisfiable so is *Skolem*(α). Therefore if either α or *Skolem*(α) is a contradiction so is the other.

Herbrand[4] *Universe* of the wff we are considering.

When we have the concept of the Herbrand Universe of a wff, we can understand the significance of *Herbrand's Theorem* on which most automated theorem-proving in general, and Prolog in particular, are based.

11.5 Herbrand's Theorem

Skolem's theorem says that a wff is satisfiable if and only if its Skolem standard form is also. Herbrand's theorem tells us that a wff is satisfiable if and only if its Skolem transform is true in *one of its Herbrand Intepretations*. The domain of a Herbrand interpretation is called its Herbrand Universe. So what is the Herbrand Universe of a wff?

The Herbrand Universe HU(S) of a Set S of Clauses

The Herbrand Universe HU(S) of a wff of in Skolem form — in the form of a conjunction, or set S, of clauses — is defined recursively. It is

(1) the set of ground terms — constants, arbitrary names, and function-symbols whose places are filled with ground terms — that appear in a set S of clauses, together with

(2) all the ground terms that may be built out of the function-symbols that occur in S, and the terms (already) in HU(S).

Example

Consider the following.

If S ={P(c)}, then HU(S) = {c}.

If S ={P(c,f(d))}, then

HU(S) = {c,d} ∪ {f(c),f(d)} ∪ {f(f(c)),f(f(d))} ...
= {c,d,f(c),f(d),f(f(c)),f(f(d)),...}

If S ={P(x), Q(f(x,d))}, then

HU(S) = {d} ∪ {f(d)} ∪ {f(f(d))} ...

and so on.

---

[4] After Jacques Herbrand, twentieth-century French logician.

Given the Herbrand Universe HU(S) of a set S of clauses — which is the domain of all possible Herbrand Interpretations of the set S of clauses — how can we pick out the Herbrand Interpretations themselves?

First we need the idea of the Herbrand *Base* HB(S) of a set S of clauses.

The Herbrand Base HB(S) of a set S of clauses

Given the Herbrand Universe HU(S) of a set of clauses S, we can go further and extract all the predicate-symbols from S and build all possible combinations of the predicate symbols in S using as terms all the terms in HU(S). This set is called the Herbrand Base HB(S).

An interpretation of S which has as its domain HU(S) associates any element of HB(S) with **T** or **F**. Collecting together all the elements of HB(S) which are assigned **T** in the interpretation into a set, we can identify that set with a Herbrand Interpretation of S.

The important result is *Herbrand's Theorem* which is this:

(1) a *set S of clauses is satisfiable iff it is true in one of its Herbrand interpretations*, or equivalently

(2) *a set S of clauses is unsatisfiable iff it has a finite set of ground (that is, variable-free) instances of its clauses is unsatisfiable.*

So Herbrand's Theorem gives us a recipe for showing a wff α to be valid, if it is valid. We negate the wff yielding $\neg\alpha$, find a Skolem standard form, represent this Skolem standard form *Skolem*($\neg\alpha$) as a set of clauses. Then we search for an unsatisfiable (finite) subset of S(*Skolem*($\neg\alpha$)).

Example

Suppose we want to show that

$$(\exists x)P(x) \vee (\forall y)\neg P(y)$$

is valid.

First we negate it, yielding

$$\neg((\exists x)P(x) \vee (\forall y)\neg P(y))$$

which is equivalent to

$$(\forall x)\neg P(x) \ \& \ (\exists y)P(y)$$

Put this into prenex normal form, which might be

$$(\forall y)(\exists x)(\neg P(x) \,\&\, P(y))$$

For Skolem standard form we need a Skolem function, say $g(y)$, so that in Skolem standard form we have

$$\neg P(g(y)) \,\&\, P(y)$$

Our set S of clauses is simply $\{\neg P(g(y))\ ,\ P(y)\}$. To set up a Herbrand Universe we must choose an arbitrary name, say 'n'. Then

$$HU(S) = \{n, g(n), g(g(n)), g(g(g(n))), g(g(g(g(n)))),\ldots\}$$

Remembering that 'y' is doing the work of the universal quantifier, S is satisfiable if and only if

$$\{\neg P(g(n))\ ,\ P(n), \neg P(g(g(n)))\ ,\ P(g(n)),\ldots\}$$

is satisfiable. The first two terms have 'n' substituted for 'y', the second two terms 'g(n)' for 'y'. The contradiction is apparent after only four clauses. So S cannot be true in any Herbrand Interpretation, and by Herbrand's Theorem it is unsatisfiable.

Herbrand's theorem guarantees that if the set S of clauses is unsatisfiable, the unsatisfiability of S will be apparent after only a finite number of substitutions of elements of the Herbrand Universe have been made. If S is not unsatisfiable, the search throught the Herbrand universe will be not terminate if the Herbrand Universe is itself infinite.

The only serious problem in showing that a set of clauses is unsatisfiable is how to go about searching through the Herbrand Universe for the unsatisfiable subset of clauses. That takes us to the next chapter, which deals with resolution and unification.

Exercise 11.3

Use the method of this section to show the validity of

(a) $(\exists x)P(x) \vee \neg P(c)$

(b) $(\exists x)(\forall y)(P(y) \Rightarrow P(x))$

Summary

There is a subtle difference between a proof-procedure and a decision procedure, the latter being the stronger algorithm. We can write a program which will eventually produce a proof of a logical truth submitted to it. That is what an automatic theorem-prover is. But we cannot write a program to test whether a wff is or is not a logical truth.

If we ask an automatic theorem-prover to prove a wff which happens not to be logically true — and therefore is not a theorem — it will fail to terminate, at least in some cases. In the case of the predicate calculus we cannot write a program to tell us in exactly which cases the theorem-prover will grind on forever, since in that case we would have a decision-procedure as well as a proof-procedure.

12 Predicate calculus: resolution and unification

In chapter 11 we saw that there is a method for refuting *formulae* in the predicate calculus.

More generally, if you want to prove that a wff α follows from a set of premises Γ — that is that the sequent $\Gamma \vdash \alpha$ is valid — then transform Γ and $\neg\alpha$ into Skolem standard form, obtaining a set of clauses. From Herbrand's Theorem it follows that we can establish that $\Gamma \cup \{\neg\alpha\}$ is unsatisifiable — if it is unsatisifiable — by a *finite* search through the Herbrand Universe $HU(\Gamma \cup \{\neg\alpha\})$. If $\Gamma \cup \{\neg\alpha\}$ is unsatisifiable, then $\Gamma \vdash \alpha$ is valid.

For the automatic theorem-prover, the computational problem is how to make this search efficient. Most contemporary theorem-provers employ *the resolution rule of inference,* applicable to the propositional calculus as described in chapter 8, augmented with an algorithm for handling the quantificational aspects of the predicate calculus.

The algorithm which handles the variables which appear in the Skolem normal form of a wff is called the *unification algorithm.* So to prove that a wff is valid we attempt to find a use unification plus resolution acting on a Skolemized version of the negation of the original wff.

The key to employing resolution theorem-proving in the predicate calculus lies in eliminating variables from sets of clauses. We eliminate variables in a set of clauses by substituting other terms in such a way that the satisfiability of the set of clauses is preserved by the substitution.

The predicate calculus derives its great expressive power from its use of variables. So we must careful in eliminating variables. We must be careful how we handle the idea of *substituting* a term for a variable. Substitution is one the key ideas of this chapter and of the next. So let us begin by considering some examples.

Example

Suppose we want to prove

$$(\forall x)P(x) \vdash P(c)$$

a sequent which is clearly valid. First we Skolemise the premises, giving the set of clauses

$$\{ \{ P(x) \} \}$$

Then we negate the conclusion to give $\neg P(c)$, and Skolemise $\neg P(c)$, giving

$$\{ \{\neg P(c) \} \}.$$

Then we take the union $\{ \{ P(x) \} , \{\neg P(c) \} \}$ of these two sets of clauses, each derived from the premises and from the conclusion, and start resolving. We hope to demonstrate the unsatisfiablity of

$$(\forall x)P(x) \vdash \neg P(c)$$

by deriving the empty clause from

$$\{ \{ P(x) \}, \{\neg P(c) \} \}.$$

The trouble is that resolution has no grip on

$$\{ \{ P(x) \} , \{\neg P(c) \} \}$$

since $P(x)$ and $\neg P(c)$ are not complementary literals.

Had we tried to prove $P(c) \vdash P(c)$, we should have obtained

$$\{ \{ P(c) \}, \{\neg P(c) \} \}$$

from which $\{ \{\} \}$ is immediately derivable.

But though resolution fails to work on

$$\{ \{ P(x) \} , \{\neg P(c) \} \}$$

our original premise $(\forall x)P(x)$ is clearly logically stronger than $P(c)$, in the sense that if $(\forall x)P(x)$ is true , so must be true $P(c)$, for any 'c'. So we should try to use this logical strength in order to make $P(x)$ and$\neg P(c)$ complementary literals. We want to *unify* $P(x)$ and $P(c)$, so that resolution can get to grips with them.

Why the term *unification*? Unification of two or more terms or expressions consists in finding a *substitution* (of terms for variables which appear in the original terms or expressions) which, when applied to each term or expression, reduces them all to a single syntactic object. Unified terms are unified to a single term, unified expressions are unified to a single expression. So to understand unification, we need a clear idea of substitution, which is a trickier notion than it might first appear.

12.1 Substitutions

Remember that P(x) in the set of clauses

{ { P(x) } , {¬P(c) } }

really means 'for all x, P(x)'. So we can *unify* P(x) and P(c) by *substituting c for x in P(x)*. (Notice that when we speak of unifying two wffs, we ignore negation signs. We ignore the fact that ¬P(c) is the negation of P(c).)

We write *the result of substituting a term t for a variable y in E* as

E[t ← y]

where t is a term, y is a variable, and E is either a term, or a proposition, or a set or list of terms, expressions or wffs etc.

t ← y is called a *substitution-component*, for reasons we shall discover shortly. Meanwhile, what do we mean by E [t ← v], the result of substituting t for v in E, where the substitution [t ← v] has a single substitution-component? The following definition is adequate, if we ignore quantifiers,[1] as we can after a wff has been Skolemized.

If E is a variable,
 then if E is the variable v, E [t ← v] is t.
If E is a variable,
 then if E is not the variable v, E [t ← v] is E.
If E is a constant,
 then E [t ← v] is E.
If E is an arbitrary name,
 then E [t ← v] is E.
If E is a term of the form $f(t_1,..t_n)$,
 then E [t ← v] is $f(t_1E\ [t \leftarrow v],..t_nE\ [t \leftarrow v])$.
If E is a wff of the form $P(t_1,..t_n)$,
 then E [t ← v] is $P(t_1E\ [t \leftarrow v],..t_nE\ [t \leftarrow v])$.

[1] In chapter 13 we shall see substitution appearing throughout the quantificational rules of inference that we develop in our natural deduction system for the predicate calculus. There we will need to take into account whether or not variables are *free* or *bound*. Here all our variables are effectively free. The substitution of terms for variables arises in the predicate calculus because the predicate calculus extends the propositional calculus essentially in virtue of its use of variables. And it will turn out that we need to be very careful about how we handle substitution, as we shall see. In fact, we shall seem to make heavy weather of it. But we must, precisely because the concepts of a variable and of substituting a term for a variable are so powerful and fundamental.

If E is a set or list of expressions or terms E_i,

then E [t ← v] is the set or list of expressions or terms E_i [t ← v].

Returning to our example, we we want to use resolution on

{ { P(x) } , {¬P(c) } }

Using the substitution [c ← x], we obtain

{ { P(x) } , {¬P(c) } } [c ← x]

read as *c for x in the set of { P(x) } and {¬P(c) }*, which is

{ { P(c) } , {¬P(c) } }

from which resolution yields

{ {} }

So (∀x)P(x) ⊢ ¬P(c) has been shown to be unsatisfiable, thus proving the original sequent (∀x)P(x) ⊢ P(c).

Example

Consider a second example, the sequent,

(∀x)P(x) & (∀x)Q(x) ⊢ P(c) & Q(d)

Skolemizing the premise, negating and Skolemizing the conclusion, we get the following set of clauses (notice the change of variable)

{ { P(x) }, { Q(y) }, {¬P(c), ¬Q(d) } }

The substiution [c ← x] gives

{ { P(x) }, { Q(y) }, {¬P(c), ¬Q(d) } }[c ← x]

which is

{ { P(c) }, { Q(y) }, {¬P(c), ¬Q(d) } }

The further substitution [d ← y] gives

({ { P(x) }, { Q(y) }, {¬P(c), ¬Q(d) } } [c ← x]) [d ← y]

which is

{ { P(c) }, { Q(y) }, {¬P(c), ¬Q(d) } }[d ← y]

which in turn is

{ { P(c) }, { Q(d) }, {¬P(c), ¬Q(d) } }

After this 'compound' substitution ([c ← x] [d ← y]), we can use resolution to derive the empty clause. We resolve the first and third clauses to obtain { ¬Q(d)} , and resolve this clause with {Q(d)} to obtain { }. So $(\forall x)P(x) \& (\forall y)Q(y) \vdash P(c) \& Q(d)$ is valid.

Example

Here is a third valid sequent

$$(\forall x)P(x,x) \vdash (\forall x)(\exists y)P(x,y)$$

Skolemizing as before we obtain from $(\forall x)P(x,x)$ and $\neg(\forall x)(\exists y)P(x,y)$ — that is $(\exists x)(\forall y)\neg P(x,y)$ — the clauses

$$\{ \{P(x,x)\}, \{\neg P(d,y)\} \}$$

Substituting d for x we get

$$\{ \{P(x,x)\}, \{\neg P(d,y)\} \}[d \leftarrow x]$$

which is

$$\{ \{P(d,d)\}, \{\neg P(d,y)\} \}$$

and not quite ready for resolution. We need to go further with an additional substitution of d for y,

$$(\{ \{P(x,x)\}, \{\neg P(d,y)\} \} [d \leftarrow x]) [d \leftarrow y]$$

yielding

$$\{ \{P(d,d)\}, \{\neg P(d,y)\} \} [d \leftarrow y]$$

which is

$$\{ \{P(d,d)\}, \{\neg P(d,d)\} \}$$

finally ripe for resolution.

In the second example we used the *composition* of substitutions

$$[c \leftarrow x] [d \leftarrow y]$$

and in the third we used

$$[d \leftarrow x] [d \leftarrow y]$$

We can write a compound substitution as a *list* of substitution-components $[c \leftarrow x, d \leftarrow y]$. In fact, in general, a substitution will be of the form of a list of substitution-components like

$$[t_1 \leftarrow v_1,\ldots,t_n \leftarrow v_n]$$

where the t_i are terms simultaneously substituted for the variables v_i. v_i are required to be *distinct* variables ($1 \leq i \leq n$).

In order to use resolution theorem-proving in the predicate calculus we need to know how to *combine* compound substitutions correctly. So consider the two substitution-components $t_1 \leftarrow v_1$ and $t_2 \leftarrow v_2$, and what we might mean by the composition

$$[t_1 \leftarrow v_1] [t_2 \leftarrow v_2].$$

First, if $v_1 = v_2$, and the variables are the same we should expect to ignore the second substitution

$$[t_2 \leftarrow v_2].$$

if t_1 does not contain an occurrence of v_1 (= v_2).

This is because, for any expression E(... v...)

$$E(\ldots v \ldots)\,[a \leftarrow v]\,[\,b \leftarrow v]$$

gives

$$E(\ldots a \ldots)[\,b \leftarrow v]$$

which remains as

$$E(\ldots a \ldots)$$

when 'a' does not contain v.

Secondly, because the substitution $[t_2 \leftarrow v_2]$ is performed *after* the substitution $[t_1 \leftarrow v_1]$, the result of the substitution $[t_2 \leftarrow v_2]$ is *affected by* the result of the substitution $[t_1 \leftarrow v_1]$. In fact we can think of it as

$$[\,t_1\,[t_2 \leftarrow v_2] \leftarrow v_1\,]$$

followed by the substitution

$$[t_2 \leftarrow v_2]$$

In other words, we first consider the effect of the substitution $[t_2 \leftarrow v_2]$ on the term t_1, yielding $t_1\,[t_2 \leftarrow v_2]$, which is to be substituted for v_1. And then consider the combined substitution to be

$$[\,t_1\,[t_2 \leftarrow v_2] \leftarrow v_1, t_2 \leftarrow v_2\,]$$

However, this is subject to two conditions, the first being that if

$$t_1[t_2 \leftarrow v_2]$$

is just v_1 - that is, that the result of substituting t_2 for v_2 in t_1 is just the variable v_1 - then $t_1\,[t_2 \leftarrow v_2] \leftarrow v_1$ would have no effect , amounting to $[v_1 \leftarrow v_1]$, and we should ignore it, that is ignore the *first* substitution-component.

The second is the one above, that if $v_1 = v_2$ we ignore the *second* substitution $[t_2 \leftarrow v_2]$.

So we are led to the following requirement :

$$[t_1 \leftarrow v_1]\,[t_2 \leftarrow v_2]$$

is to mean

$$[\,t_1\,[t_2 \leftarrow v_2] \leftarrow v_1, t_2 \leftarrow v_2\,]$$

where

(1') if $t_1[t_2 \leftarrow v_2] = v_1$, the substitution $t_1[t_2 \leftarrow v_2] \leftarrow v_1$ is *deleted,* and

(2') if $v_2 = v_1$, the substitution $t_2 \leftarrow v_2$ is *deleted.*

These rather complex conditions may be extended to the compositions of substitutions, each of which consists of one *or more* substitution-component, as in

$$[s_1 \leftarrow u_1,...,s_m \leftarrow u_m][t_1 \leftarrow v_1,....,t_n \leftarrow v_n]$$

where $[s_1 \leftarrow u_1,...,s_m \leftarrow u_m]$ and $[t_1 \leftarrow v_1,....,t_n \leftarrow v_n]$ are *lists* of substitution-components and $m > 1$ and $n > 1$.

We are led to the following rather complex definition.

$$[s_1 \leftarrow u_1,...,s_m \leftarrow u_m][t_1 \leftarrow v_1,....,t_n \leftarrow v_n]$$

is the substitution

$$[s_1[t_1 \leftarrow v_1,....,t_n \leftarrow v_n] \leftarrow u_1,$$
$$...,$$
$$s_m[t_1 \leftarrow v_1,....,t_n \leftarrow v_n] \leftarrow u_m,$$
$$t_1 \leftarrow v_1,....,t_n \leftarrow v_n]$$

where

$$s_i[t_1 \leftarrow v_1,....,t_n \leftarrow v_n] \leftarrow u_i \ (1 \leq 1 \leq m)$$

is a substitution of $s_i[t_1 \leftarrow v_1,....,t_n \leftarrow v_n]$ for each u_i. The combined substitution is subject to a generalization of the previous restrictions (1') and (2'), namely

(1) if $s_i[t_1 \leftarrow v_1,....,t_n \leftarrow v_n] = u_i \ (1 \leq 1 \leq m)$, the substitution

$$s_i[t_1 \leftarrow v_1,....,t_n \leftarrow v_n] = u_i$$

is *deleted,* and

(2) if $v_j = u_i$, for any $1 \leq j \leq n$ and $1 \leq i \leq m$, then the substitution

$$t_i \leftarrow v_i$$

is also *deleted.*

So the recipe for combining two lists of substitution-components is this. Perform the *right* substitution-list on each of the terms substituted for variables in the *left* list. Then throw out those substitution-components which now take the form of a variable substituted for itself. Finally append the *right* list of the substitution-components except for those in which a term appears substituted for a variable which was substituted for in the original *left* list.

Examples

The definition of combining two lists of substitution-components reveals that for any substitution

$$s[\,]s = s \text{ and } s[\,] = s.$$

More interestingly,

(i) $[a \leftarrow x][b \leftarrow y]$
$= \quad [a \leftarrow x, b \leftarrow y]$

(ii) $[f(y) \leftarrow x, b \leftarrow y][c \leftarrow z]$
$= \quad [f(y)[c \leftarrow z] \leftarrow x, b[c \leftarrow z] \leftarrow y][c \leftarrow z]$
$= \quad [f(y) \leftarrow x, b \leftarrow y, c \leftarrow z]$

(iii) $[f(y) \leftarrow x, b \leftarrow y][c \leftarrow y]$
$= \quad [f(y)[c \leftarrow y] \leftarrow x, b[c \leftarrow y] \leftarrow y]$
(deleting $c \leftarrow y$ by (2'))
$= \quad [f(c) \leftarrow x, b \leftarrow y]$

(iv) $[y \leftarrow x][x \leftarrow y]$
$= \quad [\,]$

(v) $[y \leftarrow x, f(y,z) \leftarrow z][x \leftarrow y, c \leftarrow z]$
$= \quad [y[x \leftarrow y, c \leftarrow z] \leftarrow x,$
$f(y,z)[x \leftarrow y, c \leftarrow z] \leftarrow z, x \leftarrow y]$
(deleting $c \leftarrow z$ by (2'))
$= \quad [f(x,c) \leftarrow z, x \leftarrow y]$ (deleting $x \leftarrow x$ by (1')).

Exercise 12.1

Write Prolog clauses for **combine**/3, a predicate which takes a pair of substitution lists and combines them to form a new substitution list.

12.2 Unification

To unify a pair of terms t_1 and t_2 we must find a substitution σ — that is, a list of substitution-components — which transforms the two terms into a single terms. σ unifies t_1 and t_2 if and only if

$$t_1\sigma = t_2\,\sigma$$

More generally, to unify two expressions E_1 and E_2, which include terms as well as wffs — we must find a substitution σ which transforms the two terms into a single terms. σ unifies E_1 and E_2 if and only if

$$E_1\sigma = E_2\,\sigma$$

There are many algorithms for finding a substitution that unifies a pair — or indeed a list — of terms. Each terminates with a substitution which does the job, or terminates with failure if there is no unifying substitution. There are many algorithms for finding the *most general unifier* — the 'smallest' substitution, that which intuitively does the least substituting to achieve unification.

Let unify(E_1,E_2) denote the most general unifier substitution, if E_1 and E_2 can be unified, and denote *fail* otherwise.

(1) If E_1 and E_2 are both variables, the substitution which simply transforms one variable into the other unifies the two variables.

(2) If only *one* of E_1 and E_2 is a variable (say E_1) then the substitution

$$[E_2 \leftarrow E_1]$$

unifies the two terms (subject to the *occur check,* for which see below). Otherwise, neither E_1 nor E_2 is a variable, in which case

(3) If both are constants, then either they are unified (which will be the case if and only if $E_1 = E_2$), or they cannot be unified and so the result is *fail.* Otherwise,

(4) If either E_1 or E_2 is a function-term or a wff, then the expression trees of E_1 and E_2 must have the same function-symbol or predicate-symbol at their roots, with the same arity, and each of their corresponding arguments must be unifiable. The unifier we want is then found by combining all the unifiers in the tree. The most general unifier is found by walking around the expression trees and combining the unifiers during the walk.

Prolog works by means of a process of unification and resolution. Its pattern-matching implements a form of unification. Prolog resolves queries by matching them with facts in a database or with the head of a rule in a database. If a query is matched with the head of a rule. Prolog proceeds to attempt to satisfy the literals in the body of the rule. Unification in Prolog works by pattern-matching, and the unifier it obtains is the environment — or association of variables with values — that Prolog reports on termination. In chapter 16 we consider the question of the extent to which this process achieves the aims of programming in logic.

Since the Prolog interpreter incorporates the unification mechanism, the simplest Prolog program for unification is

```
unify(X,X):-
   X = X.
```

assuming that variables are represented by Prolog variables — beginning with an upper-case letter or underscore character, and that constants, function-symbols and predicate-symbols are represented by literals not beginning with an upper-case letter or underscore character.

Prolog does not incorporate the *occur check* — which says that E_1 and E_2 are not unifiable if either occurs in the other. Therefore in (2), above, the variable substituted for should not occur in the term being substituted for that variable. Clearly, there is no substitution which unifes **X** and **f(X).** But try the query

```
|?- X = f(X).
```

Prolog *omits the occur check* because its cost in memory processing time. It is a good exercise to re-write code for **unify**/2 in Prolog such that **unify**/2 implements an occur check.

Exercise 12.2

Write Prolog clauses for **unify**/3, which unifies a list of terms and generates the unified term and a substitution, incorporating the occur check.

Summary

In this chapter developed the idea of a substitution-component of a substitution and the idea of a substitution, which is a list of substitution-components. We showed how a resolution theorem-prover which employed unification as a method of handling variables could act as a theorem-prover for the predicate calculus.

13 Predicate calculus: natural deduction

The language of the predicate calculus is an extension of the language of the propositional calculus. The natural deduction system we use for the predicate calculus is similarly an extension of the system for the propositional calculus. We therefore need rules of inference — elimination and introduction rules — for the two quantifiers '$\forall$' and '$\exists$', making a total of four new rules. When we come to handle *identity* - a special two-place relation, which becaue it is special is counted by the purely logical ideas of the predicate calculus — we shall need two further rules, the elimination and introduction rules for '='.

The quantifier rules make much use of two special kinds of *substitution.*

The first and most straightforward is the simultaneous substitution of *all free*[1] occurrences of a variable v by a *ground term* t in the wff P, denoted as in the previous chapter by $P[t \leftarrow v]$.

The second is the simultaneous substitution of *some free* occurrences of a variable v by a *ground term* t in the wff P, denoted by $P\{t \leftarrow v\}$. But what is a ground term? It is a term with no occurrences of variables. So

(1) constants and arbitrary names are ground terms,

(2) if $t_1,...,t_n$ are ground terms and f is an n-ary function-symbol, then $f(t_1,...,t_n)$ is a ground term,

(3) there are no other ground terms.

---

[1] An occurrence of a variable 'v' in a wff is bound if and only if either

(a) it *follows* a quantifier_symbol, as in $(\forall v)P$, or

(b) it *occurs in the scope* of a quantifier $(\forall v)$ or $(\exists v)$.

The scope of '$(\forall x)$' in '$(\forall x)P$' is the *N_or_Q_wff* 'P'.

An occurrence of a variable is free iff it is not bound. A variable is free in wff or term iff it has a free occurrence in the wff or term.

We need the idea of the *sub-terms* of a ground term when formulatating the quantifier rules. Given a ground term t, the set of its sub-terms S(t) is given by

if t is a constant or an arbitrary name,
then $S(t) = \{t\}$,
else
if t is of the form $f(t_1,...,t_n)$,
then $S(t) = \{ f(t_1,...,t_n)\} \cup S(t_1) \cup ... \cup S(t_n)$

13.1 E[t ← v] where t is a ground term

What do we mean by E[t ← v] where t is a ground term? We handle the general case in which E need not be a wff, though the rules of inference for the quantifiers appeal to the case where E is a wff. The clauses for terms are much as in chapter 12. But the clause for wffs is much more complex, because we are not dealing solely with literals.
For wffs:

If E is a wff,
then if E is $(\forall v_1)Q$
then if v is the same variable as v_1, then $(\forall v_1)Q$
else $(\forall v_1)$ (Q[t ← v])
else
if E is $(\exists v_1)Q$
then if v is the same variable as v_1, then $(\exists v_1)Q$
else $(\exists v_1)$ (Q[t ← v])
else
if E is ¬Q then ¬(Q[t ← v])
else
if E is $Q_1 * Q_2$ where $* \in \{\Leftrightarrow, \Rightarrow, \&, v\}$
then $(Q_1[t \leftarrow v]) * (Q_2[t \leftarrow v])$
else
if E is of the form $P(t_1,..t_n)$,
then $P(t_1E[t \leftarrow v],..t_nE[t \leftarrow v])$.

For terms:

If E is a term of the form $f(t_1,..t_n)$,

then $E[t \leftarrow v]$ is $f(t_1E[t \leftarrow v],..t_nE[t \leftarrow v])$.

If E is a constant or arbitrary name,

then $E[t \leftarrow v]$ is E.

If E is a variable,

then if E is the variable v, $E[t \leftarrow v]$ is t.

If E is a variable,

then if E is not the variable v, $E[t \leftarrow v]$ is E.

13.2 E{t ← v} where t is a ground term

The second kind of substitution is the simultaneous substitution of *none, some, or all* free occurrences of a term by another term in a wff 'P'. We denote this by $P\{t \leftarrow v\}$ using curly instead of square brackets. We read '$\{t \leftarrow v\}$' as 'with t for some v'.

The definition of $E\{t \leftarrow v\}$ mirrors that for $E[t \leftarrow v]$ except that the condition

If E is a variable,

then if E is the variable v, $E[t \leftarrow v]$ is t.

should be re-written as

If E is a variable,

then if E is the variable v, $E[t \leftarrow v]$ is either t or v.

The notation $P\{t \leftarrow v\}$, where P is a wff, really represents a set of formulae, rather than a particular formula. For example

$$P(x,x)\{a \leftarrow x\}$$

is

$$\{P(x,x), P(a,x), P(x,a), P(a,a)\}.$$

So what are the new rules which give us the natural deduction system of the predicate calculus?

13.3 The new rules of inference

∀-*Elimination*

From the wff '(∀x)Px' we can infer 'P[t ← x]', where 't' is a ground term. Translating this idea into a rule about sequents we can say that from

$$\Gamma \Rightarrow (\forall x)P$$

infer

$$\Gamma \vdash P[t \leftarrow x]$$

for any ground term 'm'.

A simple proof of

$$(\forall x)P(x) \vdash P(n)$$

where 'n' is an arbitrary name is

Prem	**(1)**	**(∀x)P(x)**	
1	**(2)**	**P(n)**	**1∀E**

Similarly for

$$(\forall x)P(x) \vdash P(f(a,g(c)))$$

Prem	**(1)**	**(∀x)P(x)**	
1	**(2)**	**P(f(a,g(c)))**	**1∀E**

We substitute a term for the universally quantified variable *throughout* the scope of the quantifier, as in shown in the definition of E[t ← x]

Prem	**(1)**	**(∀x)(P(x)->P(x))**	
1	**(2)**	**P(n)->P(n)**	**1∀E**

The propositional calculus rules for the five logical connectives are carried over into the predicate calculus. So a standard strategy in generating a derivation of a valid sequent in the predicate calculus is to *strip away the quantifiers with the elimination rules, then use the propositional rules, and finally re-introduce the quantifiers as necessary.* Here is a derivation of

$$(\forall x)(P(x) \Rightarrow Q(x)),\ P(f(c)) \vdash Q(f(c))$$

Prem	**(1)**	**(∀x)(P(x)->Q(x))**	
Prem	**(2)**	**P(f(c))**	
1	**(3)**	**P(f(c)->Q(f(c))**	**1AE**
1,2	**(4)**	**Q(f(c))**	**3,2->E**

∀*-Introduction*

How can we recover a universally quantified formula from a wff that is not universally quantified? Clearly, we have to place tight restrictions on when this can be done. If we know that 'P(t)' is true when 't' is the name of an *arbitrary* individual — that is, 't' is an arbitrary name — we can infer that 'P' is true of any individual (because t is arbitrary), and hence that '(∀x)P(x)' is true.

∀-Introduction implements this idea and is the inverse of ∀-Elimination. Thus from

$$\Gamma \vdash P(...m...)$$

where 'm' is an arbitrary name, infer

$$\Gamma \vdash (\forall x)(P(...m...)[x \leftarrow m])$$

subject to the following stringent condition, that 't' is a ground term, *and none of the sub-terms of 't' occur in* Γ. This last and very important restriction is meant to prevent bogus derivations like

Prem	(1)	**P(m)**	
1	(2)	**(∀x)P(x)**	**1∀E{error !!}**
	(3)	**P(m)->(∀x)P(x)**	**1,2->I**

according to which it is a logical truth that if some named individual has the property 'P', everything has the property 'P'. The error occurs at line 2, because line 1, from which line 2 is inferred, is

$$P(m) \vdash P(m)$$

and 'm' occurs in P(m).

Using ∀-Introduction we can show that bound variables can be systematically replaced with other bound variables as in

Prem	(1)	(∀x)P(x)	
1	(2)	P(n)	1∀E
1	(3)	(∀y)P(y)	2∀I

∀-Introduction is enough to enable us to justify the universal quantifier part of Stage 4 of our algorithm for converting a wff of the predicate calculus to *prenex normal form.* For example

P&(∀x)Q(x) ⊢ (∀x)(P&Q(x))

Prem	(1)	P&(∀x)Q(x)	
1	(2)	P	1&E
1	(3)	(∀x)Q(x)	1&E
1	(4)	Q(n)	3AE
1	(5)	P&Q(n)	2,4&I
1	(6)	(∀x)(P&Q(x))	5∀I

Pv(∀x)Q(x) ⊢ (∀x)(P∨Q(x))

Prem	(1)	Pv(∀x)Q(x)	
Prem	(2)	P	
2	(3)	PvQ(n)	2vI
2	(4)	(∀x)(PvQ(x))	3∀I
Prem	(5)	(∀x)Q(x)	
5	(6)	Q(m)	5∀E
5	(7)	PvQ(m)	6vI
5	(8)	(∀x)(PvQ(x))	7AI
1	(9)	(∀x)(PvQ(x))	1,2,4,5,8vE

∃-*Elimination*

Of the four quantifier rules ∃-Elimination is the least transparent, just as its close relative v-Elimination is the least transparent of the propositional rules of inference. The idea is this. Suppose we have a sequent of the form

Γ ⊢ (∃x)P

Usually, though not necessarily, P will contain one or more free occurrences of 'x'. In that case assume, in the domain in which the premises in Γ are true, that the individual of which P is true is, or can be, named 'm'.

Suppose also that we can prove

$$\Gamma_1, P[m \leftarrow x] \vdash Q$$

where neither Q nor any of the new premises in Γ_1 contains an occurrence of 'm'.

Then we can infer that

$$\Gamma, \Gamma_1 \vdash Q$$

Why is this inference allowed ? Because the first sequent guarantees that given the premises Γ, 'P[m ← x]' is true for some 'm' — though we don't know which, while the second sequent quarantees that from 'P[m ← x]' and the new premises Γ_1 we can infer 'Q'. Putting the two together we derive the third sequent. Notice that the original occurrence of the existential quantifier has been *eliminated*, so this rather complex rule is called '∃-Elimination'

Here, for example, is a derivation of

$$(\exists x)(P\&Q(x)) \vdash P$$

Prem	**(1)**	**(∃x)(P&Q(x))**	
Prem	**(2)**	**P&Q(a)**	
2	**(3)**	**P**	**2&E**
1	**(4)**	**P**	**1,2,3∃E**

Notice how ∃-Elimination works. Given the existentially quantified statement at line (1), we remove the remove the existential quantifier and replace all the bound occurrences of the quantified variable by some *new* arbitrary name 'a', new in the sense in that it occurs nowhere else in the proof so far. Notice that this line (2) is a *new premise to be discharged later*. We then derive a result which depends on the premise '2' but which does not contain an occurrence of the arbitrary name, in this case 'a'. So the result, 'P' at line (3) is independent of our choice of 'a'. But the existential statement at line (1) allows some choice for 'a'. So we are justified in swapping the dependence of 'P' on the premise '2', for a dependence on '1'. Here is where ∃-Elimination operates.

Notice that we cite:

(1) - the source of the existentially quantified statement,

(2) - the line at which we remove the existential quantifier,
and (3) - the line at which we obtain the desired formula.

Notice also that 'a' does not occur in any of the other premises on which the formula 'P' at line (3) rests.

∃-*Introduction*

∃-Introduction, by contrast, is as simple as ∀-Elimination. From

$$\Gamma \vdash P$$

we can infer

$$\Gamma \vdash (\exists x)\, Q$$

where $Q\{t \leftarrow x\}$ is P and where 't' is any ground term. (It is important to notice that not *all* the occurrences of 't' need be replaced by 'x'.)

∃-Introduction throws away information in the sense that we infer merely that something has a property from the fact that a particular thing has that property. But using ∃-Elimination and ∃-Introduction together we can show that bound variables can be systematically replaced, as for the universal quantifier.

Prem	(1)	**(∃x)P(x)**	
Prem	(2)	**P(a)**	
2	(3)	**(∃y)P(y)**	**2∃I**
1	(4)	**(∃y)P(y)**	**1,2,3∃E**

We can also justify the rest of Stage 4 of the algorithm for converting a formula of the predicate calculus to prenex normal form.

Prem	(1)	**P&(∃x)Q(x)**	
1	(2)	**P**	**1&E**
1	(3)	**(∃x)Q(x)**	**1&E**
Prem	(4)	**Q(a)**	
1,4	(5)	**P&Q(a)**	**2,4&I**
1,4	(6)	**(∃x)(P&Q(x))**	**5∃I**
1	(7)	**(∃x)(P&Q(x))**	**3,4,6∃E**

and

Prem	(1)	Pv(∃x)Q(x)	
Prem	(2)	P	
2	(3)	PvQ(a)	2vI
2	(4)	(∃x)(PvQ(x))	3∃I
Prem	(5)	(∃x)Q(x)	
Prem	(6)	Q(b)	
6	(7)	PvQ(b)	6vI
6	(8)	(∃x)(PvQ(x))	7EI
5	(9)	(∃x)(PvQ(x))	5,6,8∃E
1	(10)	(∃x)(PvQ(x))	1,2,4,5,9vE

Exercise 13.1

Prove the following sequents:

(a) ⊢ (∀x)P(f(x)) ⇒ (∃y)P(y)

(b) ⊢ P(a) ⇔ (∃y)(a = y & P(y))

(c) ⊢ (∀x)(P(x) ⇒ Q(x)) & (∃y)¬Q(y) ⇒(∃z)¬P(z)

13.4 Some valid predicate calculus sequents

Here are derivations of some useful valid sequents of the predicate calculus.

The first two tell us that we can switch around adjacent quantifiers of the same type. The third tells us that we can switch adjacent existential and universal quantifiers, if the existential quantifier is on the left. If it is on the right, we cannot switch them about. The fourth sequent shows that we must be careful when we make a quantifier the dominant operator of a wff whose dominant connective is '⇒'. The quantifier changes type — existential to universal and vice versa — if the quantifier is in the antecedent. The fifth sequent tells us a fundamental fact about symmetric, transitive relations on a given domain. If every member of the domain is related to some element of the domain, then the relation is reflexive.

Prem	(1)	(∀x)(∀y)R(x,y)	
1	(2)	(∀y)R(a,y)	1∀E
1	(3)	R(a,b)	2∀E
1	(4)	(∀x)R(x,b)	3∀I
1	(5)	(∀y)(∀x)R(x,y)	4∀I

Prem	(1)	(∃x)(∃y)R(x,y)	
Prem	(2)	(∃y)R(a,y)	
Prem	(3)	R(a,b)	
3	(4)	(∃x)R(x,b)	3∃I
3	(5)	(∃y)(∃x)R(x,y)	4∃I
2	(6)	(∃y)(∃x)R(x,y)	2,3,5∃E
1	(7)	(∃y)(∃x)R(x,y)	1,2,6∃E

Prem	(1)	(∃x)(∀y)R(x,y)	
Prem	(2)	(∀y)R(a,y)	
2	(3)	R(a,b)	2∀E
2	(4)	(∃x)R(x,b)	3∃I
2	(5)	(∀y)(∃x)R(x,y)	4∀I

Prem	(1)	(∃x)P(x)->Q	
Prem	(2)	P(a)	
2	(3)	(∃x)P(x)	2∃I
1,2	(4)	Q	1,3->E
1	(5)	P(a)->Q	2,4->I
1	(6)	(∀x)(P(x)->Q)	5∀I

Prem	(1)	(∀x)(∀y)(R(x,y)->R(y,x))	
Prem	(2)	(∀x)(∀y)(∀z)(R(x,y)&R(y,z)->R(x,z))	
Prem	(3)	(∀x)(∃y)R(x,y)	
3	(4)	(∃y)R(a,y)	3∀E
Prem	(5)	R(a,b)	
1	(6)	(∀y)(R(a,y)->R(y,a))	1∀E
1	(7)	R(a,b)->R(b,a)	6∀E
1,5	(8)	R(b,a)	7,5->E

1,5	(9)	R(a,b)&R(b,a)	5,8&I
2	(10)	(∀y)(∀z)(R(a,y)&R(y,z)->R(a,z))	2∀E
2	(11)	(∀z)(R(a,b)&R(b,z)->R(a,z))	10∀E
2	(12)	R(a,b)&R(b,a)->R(a,a)	11∀E
1,2,5	(13)	R(a,a)	12,9->E
1,2,3	(14)	R(a,a)	4,5,13∃E
1,2,3	(15)	(∀x)R(x,x)	14∀I

Prem	(1)	(∀x)Px	
Prem	(2)	(∃x)¬Px	
Prem	(3)	¬Pa	
1	(4)	Pa	1∀E
1,3	(5)	Pa&¬Pa	3,4&I
3	(6)	¬(∀x)Px	1,5¬I
2	(7)	¬(∀x)Px	2,3,6∃E
1,2	(8)	(∀x)Px&¬(∀x)Px	1,7&I
1	(9)	¬(∃x)¬Px	2,8¬I

Prem	(1)	¬(∃x)¬Px	
Prem	(2)	¬Pa	
2	(3)	(∃x)¬Px	2∃I
1,2	(4)	(∃x)¬Px&¬(∃x)¬Px	1,3&I
1	(5)	¬¬Pa	2,4¬I
1	(6)	Pa	5¬E
1	(7)	(∀x)Px	6∀I

Prem	(1)	(∃x)Px	
Prem	(2)	Pa	
Prem	(3)	(∀x)¬Px	
3	(4)	¬Pa	3∀E
2,3	(5)	Pa&¬Pa	2,4&I
2	(6)	¬(∀x)¬Px	3,5¬I

1	**(7)**	**$\neg(\forall x)\neg Px$**	**1,2,6$\exists$E**

Prem	**(1)**	**$\neg(\forall x)\neg Px$**	
Prem	**(2)**	**$\neg(\exists x)Px$**	
Prem	**(3)**	**Pa**	
3	**(4)**	**$(\exists x)Px$**	**3$\exists$I**
2,3	**(5)**	**$(\exists x)Px\&\neg(\exists x)Px$**	**2,4&I**
2	**(6)**	**$\neg Pa$**	**3,5$\neg$I**
2	**(7)**	**$(\forall x)\neg Px$**	**6$\forall$I**
1,2	**(8)**	**$(\forall x)\neg Px\&\neg(\forall x)\neg Px$**	**1,7&I**
1	**(9)**	**$\neg\neg(\exists x)Px$**	**2,8$\neg$I**
1	**(10)**	**$(\exists x)Px$**	**9$\neg$E**

13.5 Identity

The identity symbol '=' stands for a special relation '...the same as...', and unlike all the other relations expressible in the predicate calculus, it is written in infix form. Identity is a purely logical notion, and so the predicate calculus plus identity is a richer system of logic than the pure predicate calculus. It is called *first-order logic*, and is a logic adequate for mathematics and computing science.

Exercise 13.3

Paraphrase the following statements into the language of the predicate calculus, using constants, variables, predicate-, and function-symbols of your own choice, together with the identity-sign :

(a) *A non-zero number is a divisor of a second number if and only if there is a third number such that the second number times the third number is the original number.*

(b) *A number is prime if and only if it is divisible only by 1 and itself.*

(c) *The greatest common divisor of two natural numbers is the largest natural number which is a divisor of both numbers.*

Identity has the properties

(a) that $t = t$, for any term t; and

(b) that if $t_1 = t_2$, then whatever is true of (the element denoted by) t_1 is true of (the element denoted by) t_2, for any function f

$$P(\ldots,t_1,\ldots.) = P(\ldots,t_2,\ldots)$$

We can capture these ideas in two rules of inference.

=-Introduction

=-Introduction tells us that $\vdash t = t$, for any term t

=-Elimination

=-Elimination tells us that from

$$\Gamma \vdash P(\ldots t_1 \ldots)$$

and

$$\Gamma_1 \vdash t_1 = t_2$$

we can infer either[2]

$$\Gamma, \Gamma_1 \vdash P(\ldots t_2 \ldots)$$

or

$$\Gamma, \Gamma_1 \vdash P(\ldots t_1 \ldots)$$

Notice that in substituting the right side for the left side of an

[2] More precisely, from $\Gamma \vdash P$ and $\Gamma_1 \vdash t_1 = t_2$ we can infer either

$$\Gamma, \Gamma_1 \vdash Q\,[t_2 \leftarrow z] \quad \text{(i)}$$

or

$$\Gamma, \Gamma_1 \vdash R\,[t_1 \leftarrow z] \quad \text{(ii)}$$

where (i) $Q\{t_1 \leftarrow z\}$ is P and (ii) $R\{t_2 \leftarrow z\}$ is P for some variable z which does not occur in P. The idea for case (i) is that given P - which may contain some occurrences of t_1 - first substitute some (or all) of the t_1's by a new variable z and obtain Q, so that $Q\{t_1 \leftarrow z\}$ is P. Then substitute all occurrences of z in Q by t_2. Similarly for case (ii).

identity, one need not substitute *all* the occurrences of the left side. These two simple rules enable us to prove everything we need about identity, including its reflexivity, symmetry and transitivity.

$$\vdash (\forall x)(x=x)$$

	(1)	**a=a**	**=I**
	(2)	**(∀x)(x=x)**	**1∀I**

$$\vdash (\forall x)(\forall y)(x=y \Rightarrow y=x)$$

Prem	**(1)**	**a=b**	
	(2)	**a=a**	**=I**
1	**(3)**	**b=a**	**1,2=E**
	(4)	**a=b->b=a**	**1,3->I**
	(5)	**(∀y)(a=y->y=a)**	**4∀I**
	(6)	**(∀x)(∀y)(x=y->y=x)**	**5∀I**

$$\vdash (\forall x)(\forall y)(\forall z)(x=y \& y=z \Rightarrow x=z)$$

Prem	**(1)**	**a=b&b=c**	
1	**(2)**	**a=b**	**1&E**
1	**(3)**	**b=c**	**1&E**
1	**(4)**	**a=c**	**2,3=E**
	(5)	**a=b&b=c->a=c**	**1,4->I**
	(6)	**(∀z)(a=b&b=z->a=z)**	**5∀I**
	(7)	**(∀y)(∀z)(a=y&y=z->a=z)**	**6∀I**
	(8)	**(∀x)(∀y)(∀z)(x=y&y=z->x=z)**	**7∀I**

The proof of the symmetry of '=' leads naturally to a *derived rule* for identity: **=Comm** If we have a proof of

$$\Gamma \vdash t_1 = t_2$$

then we can quite easily construct a proof of

$$\Gamma \vdash t_2 = t_1$$

Exercise 13.4

Prove

(1) $a = b \vdash f(a) = f(b)$

(2) $\vdash (\forall x)(\exists y)\neg(x = y) \Rightarrow (\exists x)(\exists y)\neg(x = y)$

Exercise 13.5

Prove the derived rule **=Comm**: that if we have a proof of

$$\Gamma \vdash t_1 = t_2$$

then we can quite easily construct a proof of

$$\Gamma \vdash t_2 = t_1$$

The use of derived rules like **=Comm**, and an equally obvious rule =Trans, can make deriving results in the predicate calculus via natural deduction very much easier.

13.6 A longer example

Now for natural deduction in action in a simple problem which sometimes arises in computing science courses. Here is a proof that appending [] to the end of a list returns the original list. Or as a sequent

$$(\forall x)(x = \mathrm{append}([],x)),$$
$$(\forall x)(\forall y)(\forall z)(\mathrm{append}(\mathrm{cons}(x,y),z) = \mathrm{cons}(x,\mathrm{append}(y,z)))$$
$$\vdash (\forall x)(x = \mathrm{append}(x,[]))$$

As we mentioned in chapter 2, proofs of facts about recursively defined structures naturally imply inductive rules of inference. We bend the rules of syntax of first-order logic a little so that cons, append and [] can be represented by themselves. The inductive rule for lists is *List-Ind:*

From
$\Gamma \vdash$ P([]) *and* $\Gamma \vdash$ (∀x)(∀y)(P(y) ⇒ P(cons(x,y)))
infer
$\Gamma \vdash$ (∀x)P(x)

(∀x)(x = append([],x)),
(∀x)(∀y)(∀z)(append(cons(x,y),z) = cons(x,append(y,z)))
⊢ (∀x)(x = append(x,[]))

Prem	**(1)**	**(∀x)(x = append([],x))& (∀x)(∀y)(∀z)(append(cons(x,y),z)= cons(x,append(y,z)))**	
1	**(2)**	**(∀x)(x = append([],x))**	**1&E**
1	**(3)**	**[] = append([],[])**	*{base case}*
1	**(4)**	**(∀x)(∀y)(∀z)(append(cons(x,y),z)= cons(x,append(y,z)))**	**1&E**
Prem	**(5)**	**d = append(d,[])**	*{induction hypothesis}*
	(6)	**cons(e,d) = cons(e,d)**	**=I**
5	**(7)**	**cons(e,d)=cons(e,append(d,[]))**	**5,6=E**
1	**(8)**	**(∀y)(∀z)(append(cons(e,y),z)= cons(e,append(y,z)))**	**4∀E**
1	**(9)**	**(∀z)(append(cons(e,d),z)= cons(e,append(d,z)))**	**8∀E**
1	**(10)**	**append(cons(e,d),[]) = cons(e,append(d,[]))**	**9AE**
1,5	**(11)**	**append(cons(e,d),[]) = cons(e,d)**	**5,10=E**
1,5	**(12)**	**cons(e,d) = append(cons(e,d),[])**	**11=Comm**
1	**(13)**	**d = append(d,[]) -> (cons(e,d) = append(cons(e,d),[]))**	**5,12->I**

1 (14) $(\forall y)$(y = append(y,[]) ->
(cons(e,y) = append(cons(e,y),[])))
13∀I

1 (15) $(\forall x)(\forall y)$(y = append(y,[]) ->
(cons(x,y) = append(cons(x,y),[]))
14∀I

(16) $(\forall x)$(x = append(x,[])) 3,15List-Ind

Exercise 13.6

Show that from a proof of the sequent

$$\vdash \neg(\forall x)(\exists y)P(x,y)$$

one can construct a proof of the sequent

$$\vdash P(n,f(n))$$

where 'n' is an arbtrary name, and vice versa.

What is the connection between this result and the fact that a wff is unsatisfiable iff its Skolem standard form is?

Summary

In this chapter we explained the additional four rules of inference gioverning the quantifiers that the predicate calculus requires. We extended the predicate calculus to first-order logic by adding the identity predicate which carries with it two additional rules of inference.

14 Toy Pascal in Prolog

The formal logics of the previous chapters — culminating in the predicate calculus with identity — are logics which underlie mathematics an computing science. They also deal with the *most general* structure of reasoning and the most general logical relationships. But there are, in addition to classical logic, specialized logics which attempt to deal in detail with specific logical notions. Modal logics, for example, deal with our notion of necessity, with what follows from saying that something is necessarily the case.[1] Modal logics are usually extensions of classical logic, which handle the non-modal reasoning as a special case.

There are also specialized logics of correctness for computer programming languages. Such logics are tailored to a very special purpose, namely reasoning about computer programming source code, as a formal mathematical object. A logic of correctness for a computer programming language deals with what *assertions* follow from other *assertions* concerning what source code written in the language actually does. So if we are to develop a logic of correctness for a computer programming language we need to have a clear understanding of the syntax and semantics of the programming language the logic is meant to handle.

Naturally, we want programs to do what we specify them to do. We want them to be correct (with respect to their specifications). But a logic of program correctness will naturally be relativized to a given programming language. The logic of Pascal will be different from the logic of FORTRAN, because the two languages possess different constructs. Pascal possesses procedures to which one may pass parameters while FORTRAN does not. So a logic of program

[1] A modal logic might add to classical logic a special operator $\Box$ meaning *necessarily*. So the proposition it is necessarily the case that P can be rendered $\Box P$. In a modal logic one would expect to have theorems like $\Box P \Rightarrow P$. One might, or might not, expect to have $\Box P \Rightarrow \Box\Box P$ as a theorem. So modal logics are fun as well as potentially illuminating.

correctness for Pascal, unlike logic of program correctness for FORTRAN, has to handle statements consisting of procedure calls. And since a logic of program correctness deals with assertions about program texts — or source code — then it will also be relative to an appropriate language of assertions. The language of assertions will be the language in which we describe what is true and false of a program text.

In this chapter we define the syntax and semantics of a simple programming language, together with an *assertions language.* The programming language is structured and is a very small subset of Pascal, adequate for the programming of while-loop programs which do calculations with the integers. We call the language *Toy Pascal.* The assertions language will be sufficiently powerful for us to formulate relevant propositions about what is true of the state of a computer executing a program in Toy Pascal.

14.1 What is a logic of program correctness ?

What is a logic of program correctness so different from, and so much more complicated than the formal logic of the previous chapters.

The worlds of logic and mathematics are, in a sense, static worlds, worlds in which nothing happens, a fact which has much impressed philosophers from ancient times. For example, Euclidean geometry can be thought of as describing static geometrical objects, perfect eternal lines, triangles, circles, and not the imperfect lines, triangles and circles we draw on a blackboard. The propositional and predicate calculi can be thought of describing pure logical relationships. Both mathematics and logic seem to embody necessary and also eternal truths. Compare the business of doing mathematics with building software.

First of all, like a mathematical argument, a computer program clearly has a logic. A failed mathematical proof may go wrong because it contains a fallacy, a logical error. A computer program may execute correctly (with respect to its specification) most of the time, but crash for some inputs or deliver a wrong answer for others because it contains logical errors, bugs of the kind most difficult to trace and correct.

Secondly, but less significantly, computer programs deal with the same sorts of objects that are described by mathematicians — integers, real numbers, functions, perhaps sets, sequences etc. etc. Perhaps it is this fact more than any other that makes people think that computing is essentially mathematical, or even essentially numerical.

Programming certainly has mathematical aspects. It has aspects of correctness and error like those found in mathematics and so we demand, not unreasonably, a logic of programming against which we can measure the correctness of the programs we write.

We want a logic of programming that will perform some of the same functions for the programmer that mathematical logic performs for the mathematician. However, the programmer who wants to write correct programs need not rigorously prove them correct using a logic of programming any more than the practicing mathematician need formalize all his reasoning in formal logic[2]. But he will want a logic which will provide at least a standard of correctness. The computer scientist may want a deeper understanding of the logic, the syntax and semantics of programming languages, and of the relationship between the two. Just as logic has led mathematicians to a deeper understanding of the nature of mathematics, so a logic of programming can lead to a deeper understanding of what programs are.

The software engineer may want yet more. He may want to promote a programming methodology which enables him to develop provably correct programs. The methodology might consist of an elaboration of the idea of top-down design or step-wise refinement: break up your problem, top-down fashion, into a set of (possibly nested) procedures and use your logic of correctness to prove the correctness of your implementation of each and integration of the whole set. A logic of programming promises all this, a standard of correctness, a deeper understanding of what programs are, and a methodology for designing and implementing provable correct programs.

Now look at the differences between a piece of mathematics and a computer program. Like a piece of mathematics a program is expressed in a text, program source code. But unlike a piece of mathematics, a program also has a dynamic aspect, its execution in time. Computer programs prescribe processes in the electronic world. They control changing bit-patterns stored inside computer memories.

Here lies a fundamental source of tension in computer science and a fundamental source of difficulty in the writing of software that works.

[2] It so happens that nearly all the mathematics that mathematicians publish is informal in the sense that it is not expressed solely within the formalism of first-order logic. But in principle all of mathematics could be so expressed, even though human beings would find it completely unreadable in that form. If the logical rigor of a mathematical argument is called in question there is a standard against which to judge it. The importance of logic for the mathematician is thus two-fold. On the one hand formal logic provides an ultimate standard of mathematical rigour. And on the other, the study of mathematical logic has deepened our understanding of mathematical truth and of what mathematics is.

When we reason mathematically, our thoughts are sequential, and the text we write or the words we appear as sequence of marks or utterances. But though our reasoning is sequential, mathematical objects and the truths about them are not. When we say

$$x = 3, \textit{therefore } x^2 = 9$$

we are not describing anything that *happens*. We make from inference from the first proposition to the second by substituting the value of x given by the first equation into the second, and see that if the first is true then so is the second. The two occurrences of 'x' in the two propositions have the same meaning, or denotation or reference, namely the integer *three*.

More obviously, in mathematics there is nothing like the assignment statement

$$x := x + 1$$

Read as an *equation*, which it certainly is not, it is always false. Since mathematics is timeless there is no question of 'updating x'. When we encounter for the first time an assignment statement like this in a Pascal source code we may (and should) wonder what the two occurrences of 'x' denote.

So the key point is this. Conventional programming languages like Pascal, (and FORTRAN, COBOL, BASIC, C, MODULA-2, ADA etc.) are procedural or imperative. A program in an imperative language is a sequence of *statements*, or *commands*. The statements tell the computer both what to do and how to do it. The processor behaves accordingly.

A processor that executes a computer program written in an imperative programming language passes through a sequence of states and affects the contents of a memory, which itself passes through a corresponding sequence of memory states. And so the execution of an imperative program is essentially dynamic, and the program text, the source code, is unlike a piece of mathematics, even though it looks like a piece of mathematics to the unmathematical eye.

Here then, in the temporality of computational process and its reliance on the assignment statement, is the fundamental source of the difficulty of programming in imperative languages and the fundamental source of the complexity of their logics of program correctness.

Is there a remedy? Might a program be more like a piece of mathematics? There is every reason to think that programs in appropriately high level language may indeed approximate much better, if not quite exactly, to the mathematical ideal. Very high level languages, languages like Prolog for example, attempt to conceal from the programmer the processes or procedures which a computer executes. The extent to which programs written in such languages succeed in this aim of transcending their machine representations is a measure of their level as programming languages

But in practice programming languages always have both procedural (or imperative) and declarative aspects. Put rather vaguely, a programming language is declarative if specifying what one wants done is sufficient to get it done. A programming language is procedural if the programmer must specify how as well as what it is it is to achieve.

A fully declarative language which allowed one, when writing programs, to pay attention only to specifying programs and which allowed one to ignore how the computer would represent and execute them would be a mathematically ideal programming language. It would perhaps be an impossible programming language. At any rate, Prolog falls considerably short of this ideal as we shall see in chapter 16, if we don't know it already.

More importantly, declarative languages remain relatively inefficient in execution, at least when running on von Neumann machines. Any programming language is an interface between, and a compromise between, the human programmer and the underlying machine. Imperative languages, like Pascal, as opposed to declarative languages like Prolog, make a compromise which is much more considerate to the machine. They do this in the interest of efficiency of execution.

So we must continue to deal with the imperative style of programming. Our goal must be to develop to a logic of correctness which attempts to treat source code in an imperative language as a mathematical text about which one can reason. We view a program as a large mathematical expression which we reason about.

A program is an expression in a programming language in the same sense that a wff of propositional calculus is an expression of the language of the propositional calculus. We need special tools in order to handle computer programs — namely a logic of program correctness — partly because a program is a large mathematical expression, and partly because it is executable.

We begin by specifying a programming language. We give its syntax an operational semantics. Then we supply its logic.

14.2 Toy Pascal

We specify — at first relatively informally — a small fragment or subset of Pascal which dispenses with nearly all that is characteristic of Pascal. Toy Pascal has no declarations — no declarations of labels, constants, types, variables, procedures or functions. If Pascal is the archetypal block-structured language, Toy Pascal is un-Pascal-like. Toy Pascal lacks blocks, because a block is a set of declarations followed by a statement, the latter usually compound, composed of many smaller statements.

Toy Pascal does have structured statements. In fact Toy Pascal has the conditional, **IF...THEN...ELSE** and the **WHILE...DO...** loop. It has the assignment statement. It has sequences of commands. It also has the parenthesising of **BEGIN...END**. Here are some sample programs in Toy Pascal. The first sums the natural numbers from 0 to 10.

```
BEGIN
   number := 10;
   sum := 0;
   count := 0;
   WHILE NOT( count = number ) DO
         BEGIN
               count := count +1;
               sum := sum + count
         END
END
```

The second program calculates factorial 6 (which is equal to 720).

```
BEGIN
   n := 6;
   x := 0;
   f := 1;
   WHILE NOT (x = n) DO
         BEGIN
               x := x + 2;
               f := x * (x - 1) * f
         END
END
```

The final program calculates 2 to the power of 10.

```
BEGIN
   n := 10;
   x : = 2;
   k := n;
   y := 1;
   WHILE NOT (k = 0) DO
         BEGIN
               k := k - 1;
               y := y*x
         END
END
```

First we specify the abstract syntax of Toy Pascal. Then we specify its concrete syntax. Finally, we specify its semantics. This much makes designing a programming language exactly like designing a logic, only more complicated. If we want to implement the language on a computer, as we do later in the chapter, we must make some *pragmatic* decisions — for example, about how to store source code in files, how to access it, about limits on the size the identifiers and numerals the language can handle. First, syntax. What are the *syntactic domains* [3] of Toy Pascal? We intend Toy Pascal to be a simple programming language which enables us to perform calculations with integers. So first, there is a syntactic domain, the *numerals* $\mathbf{Z}$, each element of which takes an integer $\mathbb{Z}$, an element of the corresponding semantical domain, as its value.

Numerals are constants. But we need another syntactic domain which corresponds, at least superficially, to the concept of a variable. This is provided by the *identifiers* **Id**. Identifiers again take integers $\mathbb{Z}$ as their values, but the integer value of an identifier will depend on the appropriate state of the computer as it executes the appropriate program. Just how identifiers acquire their values is a matter for the semantics of Toy Pascal.

We need expression-forming operators $\mathbf{w_1}$, $\mathbf{w_2}$. We can add and subtract integers and so we need operators to name these operations. We can also compare integers for equality and order and so we need predicates which name these Boolean-valued operations. Therefore we need integer-valued and Boolean-valued expressions even in our very

[3] A domain, as far as we are concerned, is just a *set* of objects of a fixed type.

simple programming language Toy Pascal. The operators that we use to build the expressions form a third syntactic domain.

Toy Pascal has only one data type, the integers, although it needs to allow some expressions to take as values the truth-values, 'true' **T** and 'false' **F**. It has two broad syntactic domains in addition to those of the numerals, the identifier and the operators. They are that of expressions and that of commands.

Expressions **Exp** form the major syntactic domain which is built up from numerals, identifiers and constants. Semantically, expressions take values which are either integers, truth-values (or, in deviant cases, ERROR).

Commands **Cmd** form the final syntactic domain. Semantically, they have an effect. They change what we shall call the Store. Expressions are the declarative aspect of Toy Pascal, commands are the procedural aspect. Finally, a program **Prog** is a command, usually a compound command.

Toy Pascal is clearly a quite inadequate language for practical programming. However, in principle, any computable function is expressible within it. Any data structure, in principle if not in practice, is representable in the integers.

Inadequate though it is, Toy Pascal is a good starting point for a consideration of the problems of constructing and axiomatization of a programming language, and for considering how its semantics may handled and of how one might describe the relation between the logic of a programming language and it semantics.

14.3 Toy Pascal: syntax

The syntactic domains of a programming language are the sets of object types which make up the source code of a program in that language. So we have the following syntactic domains.

Prog	programs
Cmd	commands
Exp	expressions
$\mathbf{w_1}, \mathbf{w_2}$	operators
Id	identifiers
$\mathbb{Z}$	numerals

Now we specify the abstract syntax of Toy Pascal. The abstract syntax tells us what constructs there in the language but it tells us nothing about, for example, operator precedence or associativity. The ambiguity that an abstract syntax tolerates must be removed with a detailed, concrete syntax of the language.

Toy Pascal: abstract syntax

<Prog> ::= <Cmd>

<Cmd> ::= <I> := <Exp> |
<Cmd> ; <Cmd> |
BEGIN <Cmd> **END** |
IF <Exp **THEN** <Cmd> **ELSE** <Cmd> |
WHILE <Exp> **DO** <Cmd> | **SKIP**

<Exp> ::= <Num> |
<Id> |
<w_1> <Exp> |
<Exp> <w_2> <Exp>

<w_1> ::= **NOT** | **-**

<w_2> ::= **&** | **v** |+ | - | $*$ | = | < | ≤ | > | ≥

Toy Pascal has commands and expressions. In Toy Pascal, a program is a (usually compound) command. To generate a class of allowed expressions it needs a basis — the numerals and identifiers. It also needs ways of generating new expressions from old expressions, and for this purpose it is supplied with two types of expression-forming operator, a unary operator 'not', a unary minus '-', and binary operators '+' and '-' and '$*$', which generate integer-valued expressions from the same, and '=' which generates Boolean-valued expressions from integer-valued expressions.

As in the examples of chapter 4, we define identifiers to be strings which begin with a letter and thereafter consist of strings (possibly empty) of letters and digits. Numerals are strings of decimal digits whose values are also pre-defined. They are strings naming the integers that the strings are ordinarily taken as denoting.

Toy Pascal: concrete syntax

The abstract syntax we gave for <Cmd>, for example, was ambiguous. How, for example, are we to parse

WHILE <Exp> **DO** <Cmd> **;** <Cmd>

which is a perfectly acceptable compound command ? Is it a while-loop whose body is compound, or is a while-loop followed by a command ?

We must dis-ambiguate our grammar for commands. Since ';', the sequencer, should be thought of as binding *least tightly,* and hence as having the lowest precedence, the following concrete syntax for <Cmd> and for <Exp> will do.

```
<Cmd>     ::=  <Cmd_1> |
               <Cmd> ; <Cmd_1>

<Cmd_1>   ::=  <I> := <Exp> |
               WHILE <Exp> DO <Cmd_1> |
               IF <Exp> THEN <Cmd_1> ELSE <Cmd_1>
               | BEGIN <Cmd> END | SKIP

<Exp>     ::=  <Exp_1> |
               <Exp_1> = <Exp> |
               <Exp_1> < <Exp> |
               <Exp_1> ≤ <Exp> |
               <Exp_1> > <Exp> |
               <Exp_1> ≥ <Exp>

<Exp_1>   ::=  <Exp_2> |
               <Exp_2> + <Exp_1> |
               <Exp_2> - <Exp_1>

<Exp_2>   ::=  <Exp_3> |
               <Exp_3> * <Exp_2>

<Exp_3>   ::=  NOT <Exp_3> |
               - <Exp_3> |
               <Exp_4>

<Exp_4>   ::=  <Id> | ( <Exp> )
```

Exercise 14.1

Re-write the BNF definition of **expressions** for Toy Pascal in EBNF.

The BNF definitions of <Cmd> and <Exp> are easily transcribed as parsers in Prolog. For example, for <Cmd_1> and <Exp_1> we can write the following clauses:

```
command_1(Cmd) -->
    while,!,
    expression(Exp),
    do,!,
    command_1(Cmd_1),
    {Cmd = while(Exp,Cmd_1)}
    ;
    if,!,
    expression(Exp),
    then,!,
    command_1(Cmd_1),
    else,
    command_1(Cmd_2),
    {Cmd = if_else(Exp,Cmd_1,Cmd_2)}
    ;
    begin,!,
    command(Cmd),
    end
    ;
    skip,!,
    {Cmd = $SKIP}
    ;
    identifier(I),!,
    assignment_sign,!,
    expression(Exp),
    {Cmd = assign(I,Exp)}.

assignment_sign --> ":=".
```

and

```
expression_1(Exp) -->

    expression_2(Exp_1),!,
    (       plus,!,
            expression_1(Exp_2),
            {Exp = plus(Exp_1,Exp_2)}
            ;
            minus,!,
            expression_1(Exp_2),
            {Exp = minus(Exp_1,Exp_2)}
            ;
            {Exp = Exp_1}   ).

plus --> "+".
minus --> "-".
```

14.4 Toy Pascal: operational semantics

An operational semantics[4] of a programming language gives descriptions of an abstract machine which runs programs written in the language run. It supplies each program of the language with a history of its execution on the machine and tells us what effect each program in the language has on the abstract machine. In that sense an operational semantics evaluates any program of the language. We give an operational semantics for Toy Pascal, and this involves us in choosing appropriate domains for the constructs that appear in the language.

The numerals and the identifiers take values among the integers, a domain we call 'Num'. Expressions can therefore take integer values. Some expressions take Boolean-valued. Some are deviant semantically, like expressions which 'negate a numeral', and these take the value 'ERROR'. The set of values that expressions may take — we call them expressible values — is therefore the disjoint union of first, the basic

[4] Contrast an operational semantics with a denotational semantics for a programming language. A denotational semantics is more abstract than an operational semantics. It doesn't tell us about the history of a program's execution but merely what function a program computes. A denotational semantics for a programming language is like an operational semantics with all the operational aspects of an operational semantics removed. For more on denotational semantics see J Stoy, *Denational Semantics*, MIT, 1977.

expressible values, namely the integers 'Num', secondly the truth-values 'Bool' and thirdly set whose nly member is the error-value {ERROR}.

Besides expressions there are the commands. The meaning of a command is a function — a function which transforms a 'store' (or 'memory') into a new 'store'. An assignment statement, for example, *updates* a store.

So what is a *store* ? In our simple model of the semantics of Toy Pascal a store is an association of identifiers with their values, at least to a first, very good, approximation. If a command goes badly wrong, as a conditional statement might if the expression it contains does not take a Boolean value, then we say it takes the store into an 'error' state, again called 'ERROR'. The atom 'ERROR' is unlike an association of identifiers with their values, so we say that an (extended) store is the disjoint union of first, the basic store, which is a partial function associating identifiers with their values and secondly, the set {ERROR}.

Toy Pascal is a minute programming language which allows us to perform calculations with integers. Integers are stored in the store. The truth-values are not. So the storable values are the integers, 'Num'. The *semantic domains* of Toy Pascal are therefore the following.

Semantic Domains

Num = {0, 1, -1, 2, -2, ...} the integers
Bool = {**T**, **F**} the truth-values **T**, **F**
$\mathbf{E_V}$ the basic expressible values
$\mathbf{S_V}$ the basic storable values
$\mathbf{E_V}$ = Num + Bool
$\mathbf{S_V}$ = Num

Store = I $\rightarrow$ $\mathbf{S_V}$ (a partial function from I to Sv)

$\mathbf{E_{V+}} = \mathbf{E_V}$ + {ERROR} the extended expressible values

$\mathbf{Store_+}$ = **Store** + {ERROR} the extended store

If these are the semantic domains, what are the meanings of expressions and commands? Begin with expressions.

We need to know the shapes of the basic semantic functions. The value of a numeral is pre-defined, independently of the store. This tells us that 'Numeral_value' is a direct and pre-defined association of numerals with the integers they name. The value of an operation (unary or binary) is a mapping from an expressible_value or expressible_values into the expressible-values. The final result will be 'ERROR' if one of the input expressible_values is either itself 'ERROR' or is of the wrong type. The value of an expression is in general dependent on the store, unless the expression is a numeral. If the store is 'ERROR', or if the expression is semantically deviant, the final value is again 'ERROR'.

Semantic Functions

Numeral_value: N $\rightarrow$ **Num**

Op_1_value: $w_1 \rightarrow \mathbf{Ev_+} \rightarrow \mathbf{Ev_+}$

Op_2_value: $w_2 \rightarrow \mathbf{Ev_+} \times \mathbf{Ev_+} \rightarrow \mathbf{Ev_+}$

Exp_value: Exp $\rightarrow \mathbf{Store_+} \rightarrow \mathbf{Ev_+}$

Cmd_value: Cmd $\rightarrow \mathbf{Store_+} \rightarrow \mathbf{Store_+}$

Prog_value: Prog $\rightarrow \mathbf{Store_+}$

Expressions

The functions Op_1_value, Op_2_value and Exp_value are definable as follows. Note that we assume that the arguments of 'Exp_value' are expressions and stores and so 'Exp_value' is undefined for a syntactically ill-formed string which is intended to be an expression. If an expression is a numeral, its integer value is given by Numeral_value, independently of the store.

(1) Exp_value$[\![N]\!]$ s $\underline{\Delta}$ Numeral_value$[\![N]\!]$

The store 's' is either a set of pairs or is {ERROR}. If an expression is an identifier, and if it is stored, then its value is its stored value, otherwise its value is 'ERROR'.

(2) Exp_value⟦I⟧ s $\underline{\Delta}$ **if** for some w,
then <I, w> $\in$ s then w
else ERROR

The value of '(**NOT** E)' in the store s is the negation of the value of 'E' in s if 'E' takes a Boolean value in s. Otherwise, that is if 'E' is not a Boolean-valued expression in any store, of is s is 'ERROR', the value of '(**NOT** E)' is 'ERROR'.

$<w_1>$ ::= **NOT** | -

$<w_2>$ ::=|+ | - | ∗ | = | <

(3a) Exp_value⟦**NOT** E⟧ s $\underline{\Delta}$ **if** (Exp_value⟦E⟧ s) $\in$ Bool
then $\neg$ (Exp_value⟦E⟧ s)
else ERROR

(3b) Exp_value⟦ - E⟧ s $\underline{\Delta}$ **if** (Exp_value⟦E⟧ s) $\in$ Num
then - (Exp_value⟦E⟧ s)
else ERROR

The value of a compound expression in a store s is given by the values of the operator and of the operands in the store s, and is 'ERROR' if either operand is not integer-valued.

(4) Exp_value⟦E_1 w_2 E_2⟧ s $\underline{\Delta}$ **if** $v_1 \in$ Num **and** $v_2 \in$ Num
then Op_2_value⟦w_2⟧ v_1 v_2
else ERROR

where v_i = Exp_value⟦E_i⟧ s (i = 1,2).

Op_2_value⟦+⟧ is the addition function, Op_2_value⟦-⟧ is the difference function, and Op_2_value⟦∗⟧ is the multiplication function, all of which take numbers as their values when both inputs are numbers. They take the value ERROR when either input is ERROR. Similarly, Op_2_value⟦=⟧ is the equality function, which takes the Booleans as its values unless either input is ERROR in which case its value is ERROR These definitions are easily implemented in Prolog. For example, (4) dealing with '+' is given by:

```
exp_value(plus(E1,E2),Store,Value) :-

    exp_value(E1,Store,V1),
    exp_value(E2,Store,V2),
    (     integer(V1),
          integer(V2),!,
          Value is V1 + V2
          ;
          Value = error     ).
```

Exercise 14.2

Write a complete Prolog program for **exp_value**.

What then are the meanings of the commands ? A command takes a store into a new store. So the type of the function 'Cmd_value' which gives the meaning of commands is

$$\text{Cmd\_value: Cmd} \rightarrow \mathbf{Store_+} \rightarrow \mathbf{Store_+}$$

The assignment statement is the fundamental 'atomic' command, and its effect is to 'update' the store. So we need a primitive function 'update' whose type is

$$\text{update: I} \rightarrow \mathbf{E_V} \rightarrow \mathbf{Store} \rightarrow \mathbf{Store}$$

and which is such that

$$\text{update}(I, v, s) \;\underline{\Delta}\quad \textbf{if}\ (\exists\, w)\,(\langle I, w\rangle \in s)\quad \textbf{then}\ (s - \{\langle I, w\rangle\}\,) \cup \{\,\langle I, v\rangle)\,\}\quad \textbf{else}\ s \cup \{\,\langle I, v\rangle\,\}$$

which says that the effect of updating the value of the identifier 'I' with the value 'v' in store 's' this: over-ride the old value associated with 'I' if there is one, otherwise add the pair <I ,v>. More concretely, 'Cmd_value' takes the 'error' store into the 'error' store for call commands. This is the content of (0).

Exercise 14.3

Write Prolog clauses for **update/4**.

If s is not the 'error' store, clauses (1)-(6) apply. The meaning in store 's' of an assignment statement, assigning an expressible-value to an identifier naming a variable, is 'ERROR' if the value of the expression in 's' is not an integer, otherwise it is the effect of updating 'I' with the value of the expression in 's', in 's'. The meaning in store 's' of a sequence of commands is the meaning of second command in the store which results from the effect of the first command on 's'. The meaning of a parenthesized command in store 's' is the meaning of the command in store 's'. The meaning of an 'if...then...' command in store 's' is 'ERROR' unless the value of the expression in 's' is a Boolean, in which case the meaning of the command in 's' if the expression is false in 's' and is the meaning of the command C in 's' if the expression is true in 's'. The meaning of the 'if...then...else...' command in store 's' is as for the 'if...then..' command except that it is the meaning of the second command in 's' if the expression takes the value 'false' in 's'. The meaning of a while-loop in store 's' is 'ERROR' if the value of the loop expression is not Boolean in 's'. Otherwise if its value is false then meaning of the while-loop is simply 's', otherwise it is the meaning of the while-loop in the new store created by executing the loop body in 's'. Note that this is a recursive definition of the semantics of the while-loop.

(0) Cmd_value$[\![$C$]\!]$ ERROR $\underline{\Delta}$ ERROR

(1) Cmd_value$[\![$SKIP$]\!]$ s $\underline{\Delta}$ s

(2) Cmd_value$[\![$I := E$]\!]$ s $\underline{\Delta}$

if v $\in$ Num
then update(I, v, s)
else ERROR
where v = Exp_value$[\![$E$]\!]$ s

(3) Cmd_value⟦C_1; C_2⟧ s $\underline{\Delta}$
Cmd_value⟦C_2⟧ (Cmd_value⟦C_1⟧ s)

(4) Cmd_value⟦ BEGIN C END⟧ s $\underline{\Delta}$ Cmd_value⟦C⟧ s

(5) Cmd_value⟦IF E THEN C1 ELSE C2⟧ s $\underline{\Delta}$
if v in Bool then
if v then Cmd_value⟦C1⟧ s
else Cmd_value⟦C2⟧ s
else ERROR
where v = Exp_Value⟦E⟧ s

(6) Cmd_value⟦WHILE E DO C⟧ s $\underline{\Delta}$
if v in Bool **then**
if v then Cmd_value⟦WHILE E DO C⟧ u
else s
else ERROR
where v = Exp_Value⟦E⟧ s and
u = Cmd_value⟦C⟧ s

These equations are easily transcribed in Prolog. For example, here are the transcriptions of (0), (1), (2), and (6).

```
cmd_value(_,error,error).

cmd_value($SKIP,Store,Store).

cmd_value(assign(I,E),Store,New_store) :-
   exp_value(E,Store,Value),
   (      integer(Value),!,
          update(I,Value,Store,New_store)
          ;
          New_store=error        ).

cmd_value(while(E,C),Store,New_store) :-
   exp_value(E,Store,Value),
   (            not(bool(Value)),
                New_store=error
                ;
```

```
        (Value=false,
        New_store=Store
        ;
         cmd_value(C,Store,Next_store),
        cmd_value(while(E,C),Next_store,New_store)).
```

14.5 An interpreter for Toy Pascal

It is quite simple to implement an interpreter for Toy Pascal, and here we sketch how one might be written. Suppose one wants a small programming environment for Toy Pascal. One wants to be able to write Toy Pascal programs and store them as text in a text-file. We can assume that a handy editor is lying around, ready and waiting. But we might want to *run* a program written in a text-file, called (say) <prog>. We might want to *compile* it to an intermediate code and store that code in a file called <<prog>.toy>. And we might want *execute* the intermediate code directly. To do any of these things we need to be able access the text in <prog>, to parse it, and — if it is a syntactically correct Toy Pascal program — we want to be able find our its meaning under the function Cmd_value. In the environment one wants to run source-code stored in a text-file **run**/1. One might want to read in source-code from a text-file and compile it to an intermediate representation, which might be a list of (representations of) commands. One might want to execute this list of commands using **execute**/1. So the top level of the environment, coded in Prolog, might have clauses which look like this

```
     /* A PROGRAMMING ENVIRONMENT FOR Toy Pascal */

 /* run(file_name) runs Toy Pascal source code stored in <file_name>
   compile(file_name) compiles <file_name> to an intermediate code
              which is then stored in <file_name.toy>
    execute('file_name.toy') executes the intermediate code stored in
                          <file_name.toy> */
run(File) :-
   access(File,Code_List),nl,
   ( ( program(Parse_List,Code_List,[]),
   write('CONTAINS NO SYNTAX ERRORS'),nl,nl,
   write('*** Executing ***'),nl,nl,!,
   cmd_value(Parse_List,[],Store),nl,
```

```
    write('The final store is '),write(Store),write('.') )
    ;
    write('CONTAINS SYNTAX ERRORS') ).

compile(File) :-
    name(File,File_list),
    append(File_list,".toy",Code_file_list),
    name(Code_file,Code_file_list),
    access(File,Code_list),nl,
    ((program(Parse_List,Code_list,[]),
    write('CONTAINS NO SYNTAX ERRORS'),nl,nl,
    write('*** Writing to : '),
    write(Code_file),write(' ***'),nl,nl,!,
    tell(Code_file),write(Parse_List),write('.'),nl,told)
    ;
    write('CONTAINS SYNTAX ERRORS')).

execute(Code_file) :-
    seeing(Old),see(Code_file),
    read(Program),
    cmd_value(Program,[],Store),nl,nl,
    write('The final store is '),write(Store),write('.'),
    seen,seeing(Old),
    nl,nl,!.
```

Exercise 14.4

Complete the programming environment for Toy Pascal.

Summary

A logic of correctness for a computer programming language deals with what *assertions* follow from other *assertions* concerning what source code written in the language actually does. In order to develop a logic of correctness for a computer programming language we need a precise specification of the syntax and semantics of the language. We specified the syntax and semantics of a small language, Toy Pascal, and sketched how one might implement an interpreter for it in Prolog.

15 Program proving

What about the logic of correctness of Toy Pascal?

How can we develop a logic of programs in Toy Pascal, a logic which will tell us whether or not a program satisfies its specification?

Two ideas are suggested by this question. First, if we are to prove a program correct it must have a specification, and secondly, that a logic of correctness must deal with assertions which are in some sense 'about' the program. Take the first point. Unlike a wff in the propositional calculus — which is either a tautology or a contingency or a contradiction — a program in Toy Pascal or in any other programming language is not correct or incorrect, absolutely. It is correct or incorrect only with respect to its specification. If we are to prove a program correct with respect to its specification, we must expres the specification formally. Now take the second point. If correctness is a matter of the relationship between input and output, how can a logic get to grips with it? A logic deals with its data indirectly, via the medium of statements about the data. So the answer should be clear. Express the input and the output in the form of *assertions*, propositions that are either true or false, and develop a logic which, for each statement or command type, tells us about the relationship between the two.

Of what are assertions true or false? Again the answer is intuitively clear. Assertions are true or false of the states of the computer in which the program executes. These remarks connect the statements of the programming language with the states of the computer on which programs in that language are written. The medium of the connection is the set of assertions true or false of the computer states. We postpone the formal syntactic and semantic definition of the language of assertions for the time begin, but we expect that the language of assertions we include the expressive power of the expressions <Exp>.

Consider this again intuitively, but at a level of greater generality. Take a statement S of Toy Pascal. Now let the assertion 'P' be true of the state of the computer before the execution of S. Forget that for the moment we have not defined what 'state of the computer before the execution of S' means. Let S terminate in a state in which the assertion 'Q' is true. We express this relationship between 'P', 'S' and 'Q' as follows. We say that the correctness formula **{P} S {Q}** is true.

More formally, a *correctness formula*

{P} S {Q}

is valid iff whenever the execution of S terminates, having been initiated in a state of the computer in which 'P' is true, it terminates in a state of the computer in which 'Q' is true.[1]

Note the similarity between a correctness formula and a documented command in Pascal. We use curly brackets to surround the assertions 'P' and 'Q' just as we use the same to surround comments in Pascal, a notation which is meant to suggest to the software engineer that assertions should be part of the documentation of a well-written Toy Pascal program.

Assertions in the positions of 'P' and 'Q' in a true correctness formula '{P} S {Q}' are given special names. 'P' is called a *pre-condition of 'S' with respect to 'Q'*. 'Q' is called *a post-condition of 'S' with respect to 'P'*. Notice that we speak of 'a', and not 'the', pre-condition (post-condition) with respect to a statement and an assertion. It should be easy to see that if

{P} S {Q}

is true, then so is any

{P_1 } S {Q_1 }

where 'P_1' logically implies 'P', and 'Q' logically implies 'Q_1'. So that if P is a pre-condition of 'S' with respect to 'Q', so is 'P_1'. And similarly for Q_1.

15.1 Correctness formulae: examples

For an example of a valid correctness formula, return to the simple assignment statement x := (x + 1) Suppose we know that 'x = 0' is true before its execution. What will be true after its execution? The answer is clearly 'x = 1'.

[1] Note that this definition makes the correctness formula valid if 'C' does not terminate. Sometimes such a valid correctness formula is called a valid partial correctness formula. A valid total correctness formula is a valid partial correctness formula in which termination is guaranteed. The fact is that proving non-termination can be very difficult, even for programs written in Toy Pascal. Partial correctness — for us, correctness pure and simple — is much easier to handle in logical terms.

So

$$\{x = 0\}\ x := (x + 1)\ \{x = 1\}$$

is a valid correctness formula. In view of our previous remark, any substituting for {x = 1} any post-condition {Q} such that 'x = 1' entails 'Q', yields another true correctness formula. So we have

$$\{x = 0\}\ x := (x + 1)\ \{x > 0\}$$

is valid, since 'x = 1' entails 'x > 0'. Suppose the pre-condition had been given as 'x ≥ 0' rather than as 'x = 0'. Since the effect of the assignment statement S is to increase the value stored in the variable 'x' a valid correctness formula is

$$\{x \geq 0\}\ x := (x + 1)\ \{x \geq 0\}$$

If x were greater than or equal to zero before the execution of an assignment statement that increases its value, then it is also greater than or equal to zero after its execution. But the post-condition 'x ≥ 0' (with respect to the assertion 'x ≥ 0' and the statement 'x := (x + 1)') throws away some information. For x cannot be equal to zero after execution of the assignment statement. Clearly a stronger post-condition, stronger in the sense of more informative, is 'x > 0'. So a more informative valid correctness formula is

$$\{x \geq 0\}\ x := (x + 1)\ \{x > 0\}$$

So we have the intuitive idea that some post-conditions are stronger than others. For an assertion 'P' and a statement 'S', if both **{P} S {Q_1}** and **{P} S {Q_2}** are true correctness formulae, and if **Q_2** entails **Q_1**, then **Q_2** is a stronger post-condition than **Q_1** (with respect to the given assertion and statement). If some post-conditions are stronger than others, then is there a *strongest* post-condition, and if there is, how can we determine what it is? If **{P} S {Q}** is a true correctness formula, and if for all **R** such that **{P} S {R}** is a true correctness formula, then **Q** entails **R**, then **Q** is a *strongest post-condition* (with respect to **P** and **S**). Notice again that we are careful and say 'a' strongest post-condition. Any different post-condition *logically equivalent* to a given strongest post-condition is another strongest post-condition.

In our example of a true correctness formula

$$\{x \geq 0\}\ x := (x + 1)\ \{x > 0\}$$

it so happens that 'x > 0' is a strongest post-condition with respect to the assertion which is its pre-condition and the assignment statement. Soon we develop methods for determining weakest pre-condtitions and strongest post-conditions for the assignment statement, the compound command, and the conditional command.

Corresponding to the ideas of stronger and strongest post-conditions, we develop the ideas of weaker and weakest pre-conditions.[2] In fact, pre-conditions are, in a sense, more important than post-conditions. Although the execution of a program moves through states in which first pre-conditions are true to states in which post-conditions are true, the design process of the program usually begins with a specification of what the program is to achieve, and so begins with the post-condition and asks for the appropriate pre-condition. In practice, we find that we need a method of finding a pre-condition given a post-condition more often than we need the reverse. The idea of the weakest pre-condition with respect to a command and an assertion is the dual — or the inverse — of the idea of the strongest post-condition. Given the post-condition 'x > 0' and the statement 'x := (x + 1)', a pre-condition is clearly 'x ≥ 0'. In fact, 'x ≥ 0' is a weakest pre-condition with respect to 'x := (x + 1)' and 'x > 0'. If **{P} S {Q}** is a valid correctness formula, and for all **R** such that **{R} S {Q}** is a true correctness formula, then **R** entails **P**, then **P** is a *weakest pre-condition* (with respect to **S** and **Q**).

Notice that we are relying on our intuitions. Nothing has as yet been defined formally. We assume we know what 'assertions' are. But what exactly is an *assertion?* Let us say that any <Exp> of the programming language whose logic we are describing is an assertion.[3] We also want truth-functional compounds of assertions to be assertions, for which it is sufficient that the negation of an assertion and the conjunction and disjunction of two assertions are each assertions. However, it would be very inconvenient to represent conditional assertions using disjunction and negation, so we extend the language of assertions with a built-in conditional connective '⇒' So, on the basis of our class of expressions <Exp> we can define the syntactic class of assertions <Ass> as follows.

---

[2] The standard definition of weakest pre-condition is this. *The weakest pre-condition with respect to a programming language statement 'S' and an assertion 'Q' is an assertion 'P' which is true of all the computer states CP such that execution of S in any member of CP terminates in a finite time in a computer state of which 'Q' is true.* Notice that termination is required. What the 'computer states' are will become apparent when we define the semantics of the programming language whose statements are mentioned in the definition. We are using the expression *weakest pre-condition* in such a way as not to demand termination.

[3] We can allow non-Boolean valued expressions like 'x + 1' to be assertions even though they are neither true nor false. They simply never perform a role within the logic fo program correctness.

15.2 The language of assertions

<Assertion> ::= <Assertion_1> |
<Assertion_1> ⇒ <Assertion>

<Assertion_1> ::= <Assertion_2> |
<Assertion_2>∨<Assertion_1>

<Assertion_2> ::= <Assertion_3> |
<Assertion_3> & <Assertion_2>

<Assertion_3> ::= <Assertion_4> |
NOT <Assertion_3>

<Assertion_4> ::= **TRUE** | **FALSE** | <Exp> | (<Assertion>)

For example, given that 'x' and 'y' are identifiers,

(x = 1)

FALSE

(((NOT (x < y)) & TRUE) ⇒ (z = 0))

and

((x + 1) & FALSE)

are all assertions. If 'x' and 'y' takes values in the computer store, each, except the last, will take a boolean value and can be expected to appear in a correctness formula.

Our logic of Toy Pascal has the appearance of a natural deduction system. For each statement type, from the assignment statement to the loop, there is a rule of inference which allows one to infer a correctness formula from other correctness formulae. There is, in addition, a rule called the *rule of consequence*, which enables us to use logical truths and truths about the data structures which our programming language manipulates. We call these given truths implications.

Each instance of a rule of inference in Toy_Logic has its premises which will be already proved correctness formulae and possibly implications. The correctness formula we infer from the premises, using the rule, we call the conclusion of the rule. The rule of consequence employs implications and so, we shall see, does the rule for the if...then... conditional.

The assignment statement has a rule of inference which allows one to infer a correctness formula from no correctness formulae at all, and so may be described as an axiom schema. A proof in Toy_Logic has a tree-like structure whose leaves are instances of the axiom schema for the assignment statement, a fact which should not be surprising since any program in Toy Pascal is built up from assignment statements, which along with SKIP, form the only 'atomic' commands in the language.

But we want to prove real *programs* correct. So we develop a style of program proof which makes assertions a form of program documentation.

Exercise 15.1

Give the semantics of the assertions language. That is, define the function

$$\text{Ass\_value: Ass} \rightarrow \textbf{Store}_+ \rightarrow \textbf{Ev}_+$$

15.3 The axiom schema for the assignment statement

We begin the Toy_Logic with the axiom schema for the assignment statement which is simply this: given a post-condition 'Q' and an assignment statement 'I := E', where 'I' is an identifier (as it must be in Toy Pascal) and 'E' is an expression, the *axiom schema* is

$$\{ Q[E \leftarrow I] \} \; I := E \; \{ Q \}$$

'Q [E ← I]' means the assertion which results from substituting all the free occurrences of 'I' by 'E'. Since all the occurrences of the identifiers in assertions are clearly free, this means the strings which results from substituting all the occurrences of 'I' by 'E'.

Examples

Thus

$$(x > 0)\ [(x + 1) \leftarrow x] \text{ is } ((x + 1) > 0)$$

$$((x > 0)\ \&\ \text{TRUE})[(x + 1) \leftarrow x] \text{ is } (((x + 1) > 0)\ \&\ \text{TRUE})$$

$$(x > 0)\ [z \leftarrow y] \text{ is } (x > 0)$$

etc. For example, given the statement x := x + 1 and the post-condition {x = 1}, the axiom schema for the assignment statement tells us that since

$$\{\ x + 1 = 1\}\ x := x + 1\ \{x = 1\}$$

is valid, the pre-condition must be { x + 1 = 1}. Other instances of the axiom schema for the assignment statement are

$$\{\ 0 = 0\}\ x := 0\ \{x = 0\}$$

$$\{\ \text{NOT}\ (0 = 0)\}\ x := 0\ \{\text{NOT}(x = 0)\}$$

$$\{y = z\}\ x := y + z\ \{y = z\}$$

15.4 The axiom schema for SKIP

The axiom schema for SKIP is the simplest of all the axioms and rules. Since SKIP does nothing, must have that for any P

$$\{P\}\ \textbf{SKIP}\ \{P\}$$

15.5 The rule of consequence

The axiom schema tells that

$$\{x + 1 = 1\}\ x := (x + 1)\ \{x = 1\}$$

is a provable correctness formula, since

$$(x = 1)\ [(x + 1) \leftarrow x] \text{ is } (x + 1 = 1)$$

We should be able, in Toy_Logic, to infer from this our example

$$\{x = 0\}\ x := (x + 1)\ \{x = 1\}$$

because 'x = 0' is equivalent to '(x + 1) = 1'. So we give ourselves the powerful rule of consequence which says that for any assertions P, P_1, Q, and Q_1

$$\frac{P \Rightarrow P_1,\ \{P_1\}\ C\ \{Q_1\},\ Q_1 \Rightarrow Q}{\{P\}\ C\ \{Q\}}$$

where 'P => Q' means that 'if P then Q' is logically valid or is a truth of the data structure we are dealing with, in our case arithmetic.

Example

We apply the following *instantiation* of the rule of consequence.

$$\frac{\{x = 0\} \Rightarrow \{x + 1 = 1\},\quad \{x + 1 = 1\}\ x := (x + 1)\ \{x = 1\},\quad \{x = 1\} \Rightarrow \{x = 1\}}{\{x = 0\}\ x := (x + 1)\ \{x = 1\}}$$

Since either $P \Rightarrow P_1$ or $Q_1 \Rightarrow Q$ may be valid, we can use the following two special cases of the rule of consequence

$$\frac{P \Rightarrow P_1,\ \{P_1\}\ C\ \{Q\}}{\{P\}\ C\ \{Q\}}$$

and

$$\frac{\{P\}\ C\ \{Q_1\},\ Q_1 \Rightarrow Q}{\{P\}\ C\ \{Q\}}$$

15.6 The rule of inference for sequenced statements

The rule for sequences statements is very simple. If

$$\{P\}\ C_1\ \{Q\}$$

and

$$\{Q\}\ C_2\ \{R\}$$

are valid (and we assume, provable) correctness formulae, then

$$\{P\}\ C_1\ ;\ C_2\ \{R\}$$

is clearly also a valid correctness formula. The rule of inference for sequenced statements ensures that it is provable as well as valid.

The rule for sequenced statements is therefore

$$\frac{\{P\}\ C_1\ \{Q\},\ \{Q\}\ C_2\ \{R\}}{\{P\}\ C_1;\ C_2\ \{R\}}$$

Example

What is the pre-condition {P} in the following case?

{P}
t := x;
x := y;
y := t
{ x = Y, y = X}

Using the axiom schema for the assignment statement and the rule of consequence we can document this code as follows. First

{P}
t := x;
x := y;
{ x = Y & t = X}
y := t
{ x = Y & y = X}

then

{P}
t := x;
{ y = Y & t = X}
x := y;
{ x = Y & t = X}
y := t
{ x = Y & y = X}

and finally

{ y = Y & x= X}
t := x;
{ y = Y & t = X}
x := y;
{ x = Y & t = X}
y := t
{ x = Y & y = X}

so that we have proved that this code 'swaps' x and y.

15.7 The rule for parenthesized commands

The ';' binds least strongly of all the command-forming operators. In

Pascal and Toy Pascal, the reserved words 'begin' and 'end' serve merely as parentheses, dis-ambiguating commands like

WHILE (x = 0) DO x := 1; y := x ; x := 1

But BEGIN and END, though they structure sequences of commands, have no effect in themselves. Therefore the appropriate rule for parenthesized commands is

$$\frac{\{P\}\ C\ \{Q\}}{\{P\}\ \textbf{BEGIN}\ C\ \textbf{END}\ \{Q\}}$$

15.8 The rule for the conditional

Consider the conditional

IF E **THEN** C_1 **ELSE** C_2

If 'E' is true then 'C_1' will be executed. So that if

{P &E} C_1 {Q} [*]

is a valid correctness formula, then it must be the case that

{P} **IF** E **THEN** C_1 **ELSE** C_2 {Q}

will be a true correctness formula if 'E' is true. For in that case the conditional is just equivalent to 'C_1', and we can weaken the pre-condition since we have implicitly assumed that 'E' is true.

But what if 'E' is false, or equivalently, '(NOT E)' is true. In that case the conditional will be equivalent to 'C_2'. And by a similar argument, if

{P &(NOT E)} C_2 {Q} [**]

is a valid correctness formula, then it must be the case that

{P} **IF** E **THEN** C_1 **ELSE** C_2 {Q}

will be a true correctness formula if 'E' is false.

Since the truth and the falsity of 'E' exhaust all the possibilities, it must always be the case that

{P} **IF** E **THEN** C_1 **ELSE** C_2 {Q}

is valid, if [*] and [**] are valid.

Therefore, for the **IF ...THEN...ELSE...**statement a sound rule is

$$\frac{\{P\ \&\ E\}\ C_1\ \{Q\},\ \{P\ \&\ (\text{NOT}\ E)\}\ C_2\ \{Q\}}{\{P\}\ \textbf{IF}\ E\ \textbf{THEN}\ C_1\ \textbf{ELSE}\ C_2\ \{Q\}}$$

Example

Let us prove that the code

$$\textbf{IF } x < y \textbf{ THEN } r := y \textbf{ ELSE } r := x$$

assigns to r the maximum of x and y. In other words, let us prove that

$$\{x = X \ \& \ y = Y\}$$
$$\textbf{IF } x < y \textbf{ THEN } r := y \textbf{ ELSE } r := x$$
$$\{ (r = X \vee r = Y) \ \& \ \text{NOT}(r < X \vee r < Y)\}$$

We must prove

$$\{x = X \ \& \ y = Y \ \& \ x < y\}$$
$$r := y$$
$$\{ (r = X \vee r = Y) \ \& \ \text{NOT}(r < X \vee r < Y)\}$$

and

$$\{x = X \ \& \ y = Y \ \& \ \text{NOT}(x < y)\}$$
$$r := x$$
$$\{ (r = X \vee r = Y) \ \& \ \text{NOT}(r < X \vee r < Y)\}$$

Take the former first. The axiom for the assignment statement yields

$$\{ (y = X \vee y = Y) \ \& \ \text{NOT}(y < X \vee y < Y)\}$$
$$r := y$$
$$\{ (r = X \vee r = Y) \ \& \ \text{NOT}(r < X \vee r < Y)\}$$

so it remains to prove from the rule of consequence, that

$$\{x = X \ \& \ y = Y \ \& \ (x < y)\}$$
$$=>$$
$$\{ (y = X \vee y = Y) \ \& \ \text{NOT}(y < X \vee y < Y)\}$$

It is easy to see that this holds because

(1) $y = Y$ implies that $(y = X \vee y = Y)$, and
(2) $y = Y \ \& \ (x < y)$ implies that $\text{NOT}(y < X \vee y < Y)$.

A similar argument handles the second correctness formula needed in the premises of the rule of the conditional statement.

15.9 The rule for the while-loop

It is largely from loops that procedural programs derive their power to compress their prescription of a large number of executions of commands into a small piece of program text. In that sense, loops are a very powerful programming construct. They also need to be handled with care. Nevertheless, a sound rule of inference for a while-loop is relatively easy to formulate. Consider a while-loop **WHILE** E **DO** C.

Suppose there is an assertion 'P' which is such that, if it is true before execution of 'C', will necessarily also be true after the execution of 'C', so that {P} C {P} is a valid correctness formula. P is an *invariant* of the loop, so-called because its truth-value is invariant (at least before and after each execution of the loop body).

Note incidentally that 'C' may be a sequence of commands of the form 'C1 ; C2'. We are then considering a case in which

$$\{P\}\ C_1 ; C_2\ \{P\}$$

is valid. Of course, it need not be the case the 'P' is always true during the execution of 'C', so that neither

$$\{P\}\ C_1\ \{P\}$$

nor

$$\{P\}\ C_2\ \{P\}$$

need be valid correctness formulae.

If the loop terminates, it will merely execute C a finite number n of times, where n = 0, 1, 2,... Repeated execution of 'C' will leave the truth-value of 'P' unchanged. The expression 'E' merely serves to determine N, the number of times the loop body 'C' is executed. So the following inference must be valid.

$$\frac{\{P\}\ C\ \{P\}}{\{P\}\ \textbf{WHILE}\ E\ \textbf{DO}\ C\ \{P\}}$$

But there are *stronger* validity-preserving rules which our intuitions about the meaning of the while-loop validate. For example, the conclusion of the rule '{P} **WHILE** E **DO** C {P}' will be true if the loop does not terminate. But if it *does* terminate, then we know that 'E' must be false when it does. So we can infer the following rule must also be valid

$$\frac{\{P\}\ C\ \{P\}}{\{P\}\ \textbf{WHILE}\ E\ \textbf{DO}\ C\ \{P\ \&\ (\text{NOT}\ E)\}}$$

But we find an even stronger rule for the while-loop. For if we weaken the premise that delivers the conclusion, we have a stronger rule. The correctness formula

{P & E} C {P}

is *weaker* than the correctness formula

{P} C {P}

because given '{P} C {P}' we can derive '{P & E} C {P}' using the logical truth that '(P & E) => P' together with the rule of consequence. And yet we can see that the following rule is valid.

$$\frac{\{P\ \&\ E\}\ C\ \{P\}}{\{P\}\ \textbf{WHILE}\ E\ \textbf{DO}\ C\ C\ \{P\ \&\ (\text{NOT}\ E)\}}$$

because if 'E' is true 'P & E' carries the same information as 'P'. And if 'E' is false, the loop is not executed at all, and 'P' and '(P & (NOT E)) must have the same truth-value. So this is our rule for the while-loop. It is, in fact, the *strongest* rule that we can justify intuitively.

In the rule for the while-loop the assertion 'P' is true both before and after the execution of the loop body C. If {P} C {P} is valid, then 'P' is called a loop invariant for the loop '**WHILE** E **DO** C'. (It follows from the rule of consequence that if {P & E} C {P} is valid, then {P} C {P} is valid.)

We could demonstrate the desirable properties of the soundness and the completeness of Toy_Logic only with respect to a *semantics* for correctness formulae. But correctness formulae are complex syntactical objects. When we cite {P} C {Q} we are using a complex expression whose syntax draws on both the language of assertions, through the occurrence of 'P' and 'Q', and on the underlying programing language, through the reference to 'C'. Therefore there can be no adequate semantics for correctness formulae without a semantics for the underlying programming language.

The rules we have introduced are intuitively sound with respect to the operational semantics for Toy Pascal that we described previously. And for proving correctness to our satisfaction, that is all we need. We make no claim that our rules are complete — that all correct programs in Toy Pascal can be proved correct using our rules. The investigation of such questions as the completeness of logics of program correctness is a large topic in theoretical computing science. So let us illustrate the use of Toy_logic with some sketches of proofs of real programs.

15.10 A program proved correct

Consider our stock program which sums the natural numbers to 10.

```
BEGIN
   number := 10;
   sum := 0;
   count := 0;
   WHILE NOT( count = number ) DO
         BEGIN
             count := count + 1;
             sum := sum + count
         END
END
```

which is correct if, given the pre-condition TRUE, the post-condition of the code implies sum = $\sum_{j=0}^{j=number} j$. We prove

```
{TRUE}
   BEGIN
         number := 10;
         sum := 0;
         count := 0;
         WHILE NOT( count = number ) DO
             BEGIN
                 count := count + 1;
                 sum := sum + count
             END
   END
```

$$\{ sum = \sum_{j=0}^{j=number} j\}$$

Here we hit our first problem:

$$\text{sum} = \sum_{j=0}^{j=\text{number}} j$$

is not an assertion. Our language of assertions is insufficiently expressive. However, we do know that

$$\text{sum} = \sum_{j=0}^{j=\text{number}} j = \text{number}*(\text{number} + 1)/2$$

So we can cheat a little and replace the original post-condition with

{sum = number*(number + 1)/2}

and, in effect, use the rule of consequence to establish tthe result that

$$\text{sum} = \sum_{j=0}^{j=\text{number}} j$$

Now the problem is to prove the correctness formula

```
{TRUE}
  BEGIN
         number := 10;
         sum := 0;
         count := 0;
         WHILE NOT( count = number ) DO
              BEGIN
                   count := count +1;
                   sum := sum + count
              END
  END
 {sum = number*(number + 1)/2}
```

which we proceed to do.

Step 1

We use the rule for parenthesized commands on the outermost **BEGIN...END**. The correctness formula we want to prove will be valid if and only if

```
{TRUE}
        number := 10;
        sum := 0;
        count := 0;
        WHILE NOT( count = number ) DO
            BEGIN
                count := count +1;
                sum := sum + count
            END
{sum = number*(number + 1)/2}
```

is valid.

Step 2

Now consider the three assignments statements at the beginning of the program. Using the axiom schema for the assignment statement three times and the rule for sequenced commands twice, we can infer that

```
{TRUE}
        number := 10;
{number = 10}
        sum := 0;
{number = 10 & sum = 0}
        count := 0
{number = 10 & sum = 0 & count = 0}
```

In fact {**TRUE**} is the weakest pre-condition with respect to

number := 10; sum := 0; count := 0

and

{number = 10 & sum = 0 & count = 0}.

So proving the correctness of the program comes down to proving the correctness of

```
{number = 10 & sum = 0 & count = 0}
        WHILE NOT( count = number ) DO
             BEGIN
                  count := count +1;
                  sum := sum + count
             END
   END
{sum = number*(number + 1)/2}
```

Clearly we must now use the rule for the while-loop, which is

$$\frac{\{P \& E\}\ C\ \{P\}}{\{P\}\ \textbf{WHILE}\ E\ \textbf{DO}\ C\ C\ \{P \& (\text{NOT } E)\}}$$

So we must first find an invariant P.

Step 3

If we are to use the rule for the while-loop successfully, then the post-condition {P & (NOT E)} must imply our post-condition

{sum = number*(number + 1)/2}

However E is **NOT(count = number)**.

So {P & (NOT E)} is
{sum = number*(number + 1)/2 & NOT NOT(count = number)}
or, equivalently,
{sum = number*(number + 1)/2 & (count = number)}

Both 'sum' and 'count' change from one iteration of the loop body to the next, so this suggests that our invariant should be

{sum = count*(count + 1)/2}

First we check that this is an invariant. This is equivalent to proving

```
{sum = count*(count + 1)/2}
          BEGIN
              count := count +1;
              sum := sum + count
          END
{sum = count*(count + 1)/2}
```

that is, using the rule for **BEGIN...END**

```
{sum = count*(count + 1)/2}
      count := count +1;
      sum := sum + count
{sum = count*(count + 1)/2}
```

From the axiom schema for the assignment statement, we have

```
{sum + count = count*(count + 1)/2}
      sum := sum + count
{sum = count*(count + 1)/2}
```

and again

```
{sum + count +1 = (count +1)*(count +1 + 1)/2}
      count := count +1
{sum + count = count*(count + 1)/2}
```

So now consider

{sum + count +1 = (count +1)∗(count +1 + 1)/2}

Multiplying out we have

{sum + count +1 = (count ∗count + count + 2∗count +2)/2}

which is

{sum + count +1 = (count ∗count + count)/2 + count + 1}

so

{sum = (count ∗count + count)/2 }

or, in other words

{sum = count∗(count + 1)/2}

Therefore {sum = count∗(count + 1)/2} *is* an invariant of the loop. So all we need do now is to show that {sum = count∗(count + 1)/2} is true before the loop is entered. At this point

{number = 10 & sum = 0 & count = 0}

holds, so that so does

{sum = count∗(count + 1)/2}

since both sides of the equality are true.

All this might seem very tedious, and probably impractical as a day-to-day activity during the development of code. But we should make two or three points.

First, we might reasonably be expected to produce proofs of correctness for standard procedures — as used for sorting, searching, merging etc.[5] — which are used and re-used in many different programs.

Secondly, we might reasonably be expected to produce proofs of correctness for safety-critical procedures used in the control of fly-by-wire aircraft.

But thirdly, an understanding of assertions and the logic of programs should encourage us to document our programs properly, especially with loop invariants. Here, for example, is a (somewhat over the top) documentation of our stock program.

```
{TRUE}
   BEGIN
{TRUE}
         number := 10;
{number = 10}
         sum := 0;
{number = 10 & sum = 0}
         count := 0;
{number = 10 & sum = 0 & count = 0}
{sum = count∗(count + 1)/2}
         WHILE NOT( count = number ) DO
               BEGIN
               {sum = count∗(count + 1)/2}
               {sum + count +1 = (count +1)*(count +1 + 1)/2}
                     count := count +1;
```

[5] For good accounts of program proving for just these sorts of example see RG Dromey *How to Solve It By Computer,* Prentice-Hall ,1982. A more theoretical study is *Algorithms: The Construction, proof, and Analysis of Programs* P Berlioux and P Bizard, John Wiley, 1986.

```
        {sum + count = count*(count + 1)/2}
             sum := sum + count
        {sum = count*(count + 1)/2}
        END
{sum = number*(number + 1)/2}
  END
{sum = number*(number + 1)/2}
```

Exercise 15.2

Prove the correctness of

(a)

```
{TRUE}
BEGIN
  n := 4;
  x := 0;
  f := 1;
  WHILE NOT (x = n) DO
      BEGIN
          x := x + 2;
          f := x * (x - 1) * f
      END
END
{f = n!}
```

(b)

```
{TRUE}
BEGIN
  n := 10;
  x : = 2;
  k := n;
  y := 1;
  WHILE NOT (k = 0) DO
      BEGIN
          k := k - 1;
          y := y*x
      END
```

END
$\{y = 2^n\}$

Project: Implement an interpreter for Toy Pascal which accepts assertions in source-code and which evaluates their truth-values during program execution. The program should crash abs issue an error message if an assertion is false.

Summary

In this chapter we described a logic for the small programming language Toy Pascal. We gave rules of inference for each of the constructs of the language, developed the concept of a weakest pre-condition with respect to a statement of Toy Pascal and a post-condition. We showed how small programs could be documented and proved correct by using the logic of program correctness that we set up.

16 The logic of Prolog

To the programmer familiar only with an imperative language like Pascal, logic programming in general and the programming language Prolog in particular, will appear magical. Programming in logic is an ideal to which real logic programming languages approximate. An ideal logic programming language — which Prolog is not — would allow one simply to specify the solution to a problem and would, in principle, remove the burden of implementing an algorithm which would find the solution.

Programming in conventional, imperative languages is of course quite unlike this. The programmer spends most of his time designing procedures which, called in an appropriate sequence, solve his problem. He first specifies his problem; that is, he gives a precise account of what it is he wants solved. Then he specifies how his program is to solve it. He constructs an algorithm. Most programming does seem to be a matter of choosing the best data structures and implementing the best form of flow of control. Proving the correctness of a procedural program — like proving the correctness of programs written in Toy Pascal — is not a trivial matter, as the previous chapter may have suggested to you.

Might it be possible to eliminate the need for thinking at the procedural level? Might it be possible to leave the *how* of software to a programming language? The conventional programmer may hold that[1]

Program = Algorithm + Data Structure.

The logic programmer[2] says

Algorithm = Logic + Control.

In an ideal logic programming language, the 'Control', the 'how' or imperative aspects of programming, are to be handled by the inference

[1] An idea which suggested, or perhaps comes from, the title for Niklaus Wirth's classic book *Algorithms + Data Structures = Programs,* Prentice-Hall, 1976.

[2] Because *Algorithm = Logic + Control,* the title of a classic paper by Robert Kowalski (1979). See also the discussion by Christopher Hogger in his book *Introduction to Logic Programming,* Academic, 1984 p. 99ff.

engine of the language. In actual logic programming languages, of which Prolog is by far the most widely used, the programmer will have to pay some attention to the control component, especially for non-trivial programs. Thus Robert Kowalski writes

> *Logic programs express only the logic component L of algorithms. The control component C is exercised by the program executor, either following its own autonomously determined control decisions or else following insructions provided by the programmer.*
>
> *The conceptual separation of logic from control has several advantages:*
>
> *(1) Algorithms can be constructed by successive refinement, designing the logic component before the control component.*
>
> *(2) Algorithms can be improved by improving their control component without changing the logic component at all.*
>
> *(3) Algorithms can be generated from specifications, can be verified and can be transformed into more efficient ones, without ever considering the control component, by applying deductive rules to the logic component alone.*
>
> *(4) Inexperienced programmers and database users can restrict their interaction with the computing system to the definition of the logic component, leaving the determination of the control component to the computer.*[3]

How far does Prolog fall short of Kowalski's *ideal*? How far does any logic programming language necessarily fall short of this ideal?

16.1 Prolog: declarative and procedural semantics

What does the following program mean ?

```
p(a).            /* 1 */

p(b).            /* 2 */

q(X) :- p(X).    /* 3 */
```

Clearly what a program means must determine which queries will succeed, given the program. It is easy to say what these are. The query

[3] Robert Kowalski, *Logic for Problem Solving*, North-Holland, 1979, pp.125-126.

| ?- p(X).
succeeds with **X = a** and **X = b**, and the query
| ?- p(X).
succeeds with **X = a** and **X = b**.
The appropriate conjunctions and disjunctions, like
| ?- p(X) ; q(X).
| ?- p(X) , q(Y).
will also succeed. In fact the program acts as if it contained no rules, just the facts in its smallest Herbrand Interpretation

$$H_{min} = \{p(a),p(b),q(a),q(b)\}$$

All Herbrand Interpretations which include this one satisfy the three clauses of the program, and H_{min} is the smallest of them, or the intersection of them. Exactly the same applies for more complex Prolog programs. The meaning of a Prolog program is its smallest Herbrand Interpretation. But what is the connection between this, the *declarative interpretation* which abstracts from how the Prolog interpreter works, and the *procedural interpretation*, which is based on the fact that the Prolog interpreter executes.

One can get a feel for the relationship between the two by looking at the denotational semantics of Prolog. According to the theory of logic programming[4] a logic program P can be associated with a function F_P which maps Herbrand Interpretations of P — subsets of the Herbrand Base of the program — on to Herbrand Interpretations of P. Given an Herbrand Interpretation H_n, $F_P(H_n)$ is larger, that is

$$H_n \subseteq F_P(H_n).$$

Explicitly,

$$F_P(H)= \{A :\text{-}\ A_1,..A_n \mid A :\text{-}\ A_1,..A_n \text{ is a ground instance of one of the clauses in the program P and } \{A_1,..A_n\} \subseteq H\}.$$

Facts in the program can be thought of as clauses like A :- $A_1,..A_n$ with $n = 0$.

Begin with the empty Herbrand Interpretation { }, and keep on iterating F_P and you scoop up more and more of the Herbrand Base of P. Eventually you reach a *fixed point*. For some n,

$$F_P{}^n(\{\}) = F_P{}^{n+1}(\{\})$$

and this is the least Herbrand Interpretation, which we can call H_∞, that satisfies the program P.

[4] See JW LLoyd, *Foundations of Logic Programming*, Springer-Verlag, 1984, Chapter 1.

For example, for the program above

$$Fp(\{\}) = \{p(a),p(b)\}$$

$$Fp^2(\{\}) = \{p(a),p(b),q(a),q(b)\}$$

$$Fp^3(\{\}) = \{p(a),p(b),q(a),q(b)\}$$

and so

$$Fp^\infty(\{\}) = \{p(a),p(b),q(a),q(b)\}$$

We can picture the process thus.

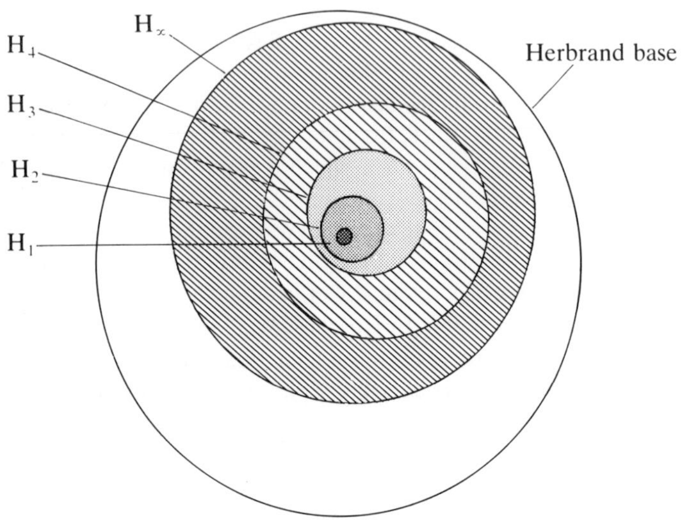

Beginning with the empty Herbrand Interpretation H_0, iterate to obtain H_1, H_2,... through to H_∞.

How well is the convergence on the smallest Herbrand Interpretation H_∞ mirrored in the procedural interpretation? The answer is that in one respect they match perfectly. In the sense that a logic programming language like Prolog matches goals against facts and the heads of rules in a program it is buiding what is called might be called a resolution tree. This tree is independent of how the logic programming language builds it. The goal will succeed when it forms the empty clause by resolution. For example, given the simple program

```
p(a)                    /* 1 */

p(b)                    /* 2 */

q(X) :- p(X).           /* 3 */
```
and the query **:- q(Y)**

Prolog succeeds with **Y = a** and then with **Y = b** by building the following resolution tree.[5]

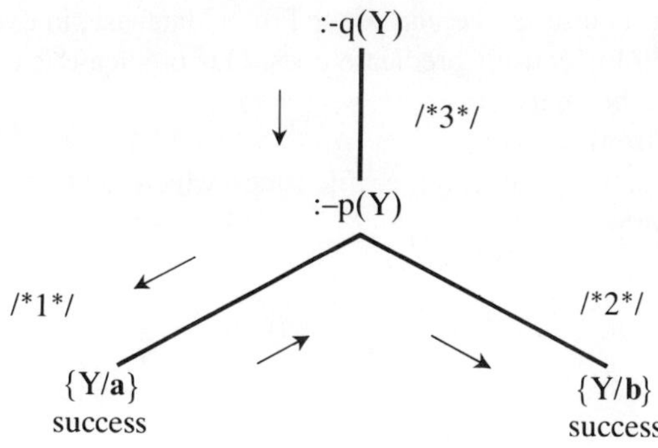

The query q(Y) unifies and resolves with clause /* 3 */, and the resultant p(Y), unifies and resolves first with clause /* 1 */, then with clause /* 2 */ to succeed with Y = a and Y = b, in that order.

Prolog searches the resolution tree from *left to right*, using a *depth-first* mechanism. We write the clauses which are resolved earlier on the left hand side of the resolution tree. So the *left to right* description mechanism amounts to a way of representing a selection rule for the clause to be resolved.

The notion of the order in which the tree is searched is important because Prolog does not first build the tree and then search it for successes. It searches the tree while building it, and the peculiarities of Prolog's behaviour in practice are largely due to the search mechanism

[5] Technically known as an *SLD-tree* because Prolog uses SLD resolution. This means L(inear resolution) on D(efinite clauses) with a S(election function). See See JW LLoyd, *Foundations of Logic Programming*, Springer-Verlag, 1984, Chapter 2.

that it employs, rather than to its underlying logical machinery of resolution and unification. Prolog uses a depth-first search strategy. This forces real Prolog to depart from its procedural interpretation. Very often, a Prolog program will fail to terminate, not because the resolution tree is infinite and there are no successes, but because it goes off searching down an infinite branch of the tree, ignoring the possibility of successes in finite branches of the tree.

16.2 Experimenting with Prolog

So let us consider how we might experiment with by, in effect, altering its search rule. We choose cheap and dirty methods. We can write, in Prolog itself, a simple loader which reads Prolog clauses from a text-file, and asserts the clauses at the *end* of the Prolog database, in exactly the way that the Prolog consult predicate does. Our predicate is called **consult_1**, so that the query

```
:-  consult_1(<filename>)
```

loads a Prolog program in the text_file <filename> which can be then be queried in the usual way.

```
/* Prolog_1 */

consult_1(Program) :-

   seeing(Current),
   see(Program),
   load_1,
   seen,
   see(Current).

load_1 :-

   read(X),
   (X = end_of_file
   ;
    assertz(X),
    load_1).
```

Suppose we want to *change* the Prolog search mechanism. Suppose we want to reverse the order in which clauses in the program in

<filename> are presented for resolution. How should we proceed? One solution is to use the built-in Prolog predicate **asserta/1**, rather than **assertz/1**, in the Prolog_loader, so that the clauses are deposited in the Prolog database in the reverse order to that in which they appear as text in the program <filename>. The clause for **load_1** should read

```
load_1 :-

    read(X),
    (X = end_of_file
     ;
     asserta(X),
     load_1).
```

The resolution tree for the program

```
p(a).                   /* 1 */
p(b)                    /* 2 */
q(X) :- p(X).               /* 3 */
```

now looks like this:

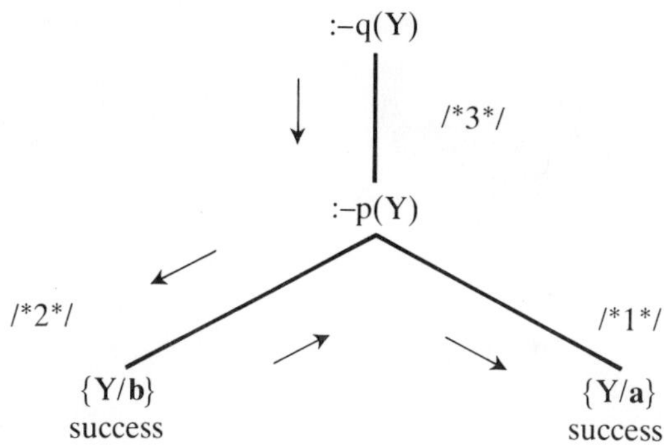

So we can easily change the order in which Prolog selects the clauses in a program. Now suppose we want to reverse the order in which the *literals* in each clause are unified and resolved. How can we achieve this outcome by modifying our loader ? Clearly we want a re-write rule which takes a clause like

```
p :- q,r,s,t
```

and translates it into

```
p :- t,s,r,q.
```

The predicate *rewrite _2* does just this.

```
/* Prolog_2 */

consult_2(Program) :-

   seeing(Current),
   see(Program),
   load_2,
   seen,
   see(Current).

load_2 :-

   read(X),
   (X = end_of_file
    ;
   rewrite_2(X,Y),
    assertz(Y),
    load_2).

rewrite_2(:-(H,B), :-(H,B1)) :-

   rewrite_2(B,B1).

rewrite_2((C1,C2), (C2,C1)).

rewrite_2(X,X).
```

The key clause is clearly

```
rewrite_2((C1,C2), (C2,C1)).
```

Now suppose we want to change the *syntax* of Prolog, so that the

usual Prolog clause
s :- (p ; q),r.
reads as
p v q & r => s.

The connectives '=>', 'v' and '&' are arranged in order of increasing precedence. We can let each be right-associative, though strictly '=>' is not associative at all since

p => q => r.

should be ill-formed, and neither 'v' nor '&' can figure in the *consequent* of a clause. It is simple to implement such a translator — which we call Prolog_3 — in Prolog.

```
                    /* Prolog_3 */


:-  op(255,xfy,'=>').
:-  op(254,xfy,'v').
:-  op(253,xfy,'&').

consult_3(Program)  :-

   seeing(Current),
   see(Program),
   load_3,
   seen,
   see(Current).

load_3  :-

   read(X),
   (X  =  end_of_file
   ;
   rewrite_3(X,Y),
    assertz(Y),
    load_3).
```

```
rewrite_3('=>'(Antecedent,Consequent) ,
    ':-'(Consequent,Antecedent1)) :-

  rewrite_3(Antecedent, Antecedent1).

rewrite_3('&'(A,B), ','(A1,B1)) :-

  rewrite_3(A,A1),
  rewrite_3(B,B1).

rewrite_3('v'(A,B), ';'(A1,B1)) :-

  rewrite_3(A,A1),
  rewrite_3(B,B1).


rewrite_3(X,X).
```

Suppose we *consult_3* a simple program like

```
p(a).
q(a).
s(X) v p(X) & q(X) => r(X).
```

we add the following clauses, in the following order, to the Prolog data-base.

```
p(a).

q(a).

r(_17) :-
    s(_17);
  (    p(_17) ,
       q(_17)      ).
```

as is shown by *listing* it. The query
:- r(Y)
succeeds with **Y = a**.

16.3 Prolog as a logic programming language

How far do Prolog and its present implementations fall short of the logic programmer's ideal? In what ways is Prolog unlike a true logic programming language? First, Prolog restricts programs to containing *Horn clauses*, clauses in which there is at most one positive literal. This has the effect of restricting the expressive power of the Prolog disjunction. A clause like

r :- (p ; q).

is allowed in a Prolog program, because it may be translated into the two clauses

r :- p.

and

r :- q.

But a clause like

(q ; r) :- p.

is not a Horn clause and is not equivalent to a collection of Horn clauses, and so Prolog would not allow it to appear in a program. The limtations on conjunction are not so drastic. Thus while

(q,r) :- p.

is not a Horn clause, because it has two positive literals, 'q' and 'r', it is equivalent to the two Horn clauses

q :- p.

and

r :- p.

Of course, literals can be conjoined in the body of a clause, each such literal being negative. If Prolog has a conjunction — the ',' — which approximates in expressive power to the conjunction of classical logic and if has a relatively inexpressive disjunction, its conditional is scarcely worthy of the name. In Prolog one cannot have a clause like

r :- (q : - p).

One cannot 'nest' the conditional and so the questions of its associativity, as a binary operator, does not arise.

Of course, in classical logic $\mathbf{P} \Rightarrow (\mathbf{Q} \Rightarrow \mathbf{R})$ and $(\mathbf{P}\ \&\ \mathbf{Q}) \Rightarrow \mathbf{R}$ are logically equivalent, so one can rewrite the offending clause as

r :- q,p.

Given these restrictions on the Prolog 'connectives' it is surprising that the translation from a program specification to Prolog code is usually so straightforward. This is not to say that there are no near impossible cases, and some of these are much discussed in the technical literature. An important reason for Prolog's falling short of being an ideal logic programming langauge is the use it makes of *depth-first search* of the resolution tree. This is what causes it to come unstuck with programs like

```
p.
p :- p.
```

and

```
p :- p.
p.
```

If the first is queried with

```
| ?- p.
```

it will succeed, though the search for more solutions will send Prolog's inference engine into a loop. The second goes into a loop immediately. Both programs are logically equivalent to the simple program

```
p.
```

But the 'tautology'

```
p :- p.
```

clearly has a devastating effect on the inference engine. Not only do logically equivalent programs behave differently, but the commutivity of the disjunction fails. Given the query

```
| ?- (p ; q).
```

the program

```
q.
p :- p.
```

goes into a loop immediately. It succeeds if supplied with the query

```
| ? - (q ; p).
```

Similar behaviour can easily be made to occur if the disjunction is allowed to appear in the program.

These points apply to Pure Prolog, supplemented with a disjunction, and implemented with the depth-first, 'left-to-right', search rule. They make it very difficult for there to a simple *logic of Prolog*, for a logic of Prolog would have to handle the failure of logically equivalent programs to behave identically, the non-commutivity of disjunction, and the lack of the expressivity of the 'conditional'.

Prolog, complete with negation, is certainly a more wieldy language than Prolog without it, but Prolog's negation is even more problematic than its disjunction or conditional. But Prolog's negation introduces more problems. The query

| ?- not p.

succeeds if and only if the query

| ?- p.

fails. And so Prolog's negation is a meta-linguistic negation meaning roughly *not provable from the database*. In some respects Prolog's negation behaves as one would expect. Thus the query

| ?- p : not p.

always succeeds, while

| ?- p, not p.

always fails.

It is fair to say that Prolog, though it is an approximation to a logic programming language, does not really contain proper logical connectives at all, and that therefore the most appropriate view to take of it is that it is a very high-level procedural programming language which has a logic programming look about it. Viewed in that way, Prolog is an excellent tool for prototyping and for writing programs that manipulate terms in exactly the way that programming logical ideas demands.

Project

Implement your own version of New-Prolog by first giving a context-free grammar for the language, secondly by writing a parser which generates a representation of, and executes queries to, a consulted text-file containing New-Prolog code.

Summary

In this chapter we described the declarative interpretation of Prolog programs and compared that interpretation with its procedural and operational interpretations. We considered various ways in which the search strategy of Prolog could be altered. Finally we considered the merits of Prolog as a means of *programming* in logic.

Glossary

axiomatic system In an axiomatic system of logic one is given a small set of axioms, wffs which are presented by fiat as primitive theorems, together with a small number of primitive rules of inference.

BNF Backus-Naur Form BNF notation is a way of representating of the production set of a context-free grammar (or type 2). Non-terminals are written as identifiers surrounded by the brackets '<' and '>'; terminals are written as themselves, or as the names of sets of terminals; and the symbol '->' is written as '::='. Productions which a fixed non-terminal of the left are written as alternatives on the right using the symbol '|' which can be read as 'or'.

Chomsky Hierarchy The Chomsky Hierarchy results from placing restrictions on the forms of the production rules that the phrase structure grammar may take. A phrase structure grammar is of type 3 if the left-hand side of each production rule is a single non-terminal, and if the right-hand side of each production rule contains at most one non-terminal symbol which must then be the right side symbol farthest to the right. A phrase structure grammar is of type 2 if the left side of each production rule is a single non-terminal symbol. A phrase structure grammar is of type 1 if the length of the right-hand side of each production rule is at least as great as the length of its left-hand side. All phrase structure grammars are trivially of type 0 by definition.

completeness A proof system for a formal logic is *complete* with respect to an interpretation of the system if every wff or sequent which is valid according to the interpretation is provable.

context-free grammar A context-free grammar is a phrase structure grammar of type 2 in the Chomsky Hierarchy.

context-free language A context-free language is a language generated by a context-free grammar.

conjunctive normal form A wff in conjunctive normal form is a conjunction of clauses; that is, a conjunction of disjunctions of literals.

clause A clause is a disjunction of literals. It may also be regarded as a *set* of literals.

Horn clause A Horn clause is a clause with at most un-negated literal. There essentially four kinds of Horn clause.
(1) the *empty* clause, containing no literals,
(2) a unit clause, containing just one un-negated (or positive) literal, and no negative literals.
(3) a clause lacking a positive literals, but containing at least one negative literal, and
(4) a clause containing exactly one positive literal, and at least one negative literal.

A Prolog program contains clauses of kinds (2) - the *facts* - and (4) - the rules. Rules are clauses of the form

$$H \vee \neg B_1 \vee \neg B_2 \vee \ldots \vee \neg B_k$$

which are written as

$$H :- B_1, B_2, \ldots, B_k$$

Queries are clauses of type (3).

decision procedure A *decision procedure* for some property ψ of a set of objects is an algorithm that determines whether or not any element of that set has ψ.

fact see *Horn clause*.

literal A literal is either an atom, or the negation of an atom.

natural deduction Natural deduction systems of logic dispense with axioms, and compensate for this loss by incorporating many more rules of inference. A natural deduction system usually has two rules for each logical connective in the language of the logic with which it deals -

one rule for *introducing* the connective, and one for *eliminating* it from wffs.

phrase structure grammar A phrase structure grammar is a meta-linguistic structure which generates a formal language. It consists of four items:

(1) a set of (meta-linguistic) variables which we allow to range over strings of the language, whether sentences or not;

(2) a set of (meta-linguistic) constants which designate the corresponding basic symbols, or 'terminals', of the language;

(3) a set of rules, called *production rules* which enable us to derive the sentences of the language; and

(4) a final special symbol, the so-called *start symbol* , for which the letter 'S' is usually reserved.

proof-procedure A proof-procedure is an algorithm for *generating* the proof a wff (say) whenever the wff is valid. A proof procedure may fail to terminate in failure if the wff is not valid.

query see *Horn clause*.

rule of inference A rule of inference is a meta-linguistic object which allows one to derive a formula, or a sequent, from other formulae or sequents.

sequent A sequent is a pair consisting of a set of premises and a conclusion, which is a single wff. In a generalized sequent, the conclusion is a *set* of wffs.

sequent calculus In a sequent calculus, the proved items are generalized sequents, objects of the form $\Gamma \vdash \Delta$ where Γ and Δ are sets (or possibly lists) of wffs. The intuitive interpretation of a provable generalized sequent $\Gamma \vdash \Delta$ is: whenever all the wffs in Γ are true, then at least one of the wffs in Δ is true, or, equivalently for finite Γ and Δ: whenever a conjunction of al the wffs in Γ is true, any disjunction of all the wffs in Δ is true. In a sequent calculus, each connective has rules for introducing both into the premise-set Γ and into the conclusion-set Δ.

soundness A proof system for a formal logic is *sound* for an interpretation of the system if every provable wff or sequent is valid according to the interpretation.

Solutions to selected exercises

Exercise 3.1 We can generalise the idea of a binary tree to the idea of a tree in general (or an n-ary tree) dropping the word 'binary' in the clauses for binary trees and by replacing clause (2) for **Bintree** by

(2') if N is a node, and if $T_1 ... T_n$ are trees, then **mk-Tree$_n$** $(N, T_1, ..., T_n)$ is a tree.

The values of the functions make_tree' (there is a different on for each n), may be represented by the lists $[N, T_1, ..., T_n]$ etc.

Exercise 4.2.

```
        /* Context-Free Grammar for the language of sets */

:- consult(make_sets).
:- consult(numerals).

set(String,Set) :-
   set_string(Set,String,[]) ,
   make_set(List,Set) .

set_string(List) -->
   left_curly_bracket,!,
   list_of_elements(List),!,
   right_curly_bracket.

list_of_elements(List) -->
   element(Element),!,
   (     comma,!,
         list_of_elements(List_1),!,
         {List = [Element | List_1]}
```

```
        ;
        {List = [Element]} )
   ;
   {List = []}.

element(Element) -->
   set_string(List_1),!,
   {Element = List_1}
   ;
   numeral(Element).

left_curly_bracket --> "{".
 right_curly_bracket --> "}".
```

Exercise 5.4.

```
<sequent>          ::=  <list_of_wffs> '|-' <list_of_wffs>
<list_of_wffs>     ::=  <empty>  | <wff>
                        | <wff> ',' <list_of_wffs>
```

```
                          /* Sequents */

:- consult(pc_wffs).

sequent(String,List) :-
   strip(String,New_String),
   make_sequent(List,New_String,[]).

make_sequent([X,Y]) -->
   list_of_wffs(X),
   turnstile,
   list_of_wffs(Y).

list_of_wffs(X) -->
   (empty, assign(X,[]) )
   ;
    (wff(Y), ( (empty, assign(X,[Y]) )
             ;
```

```
            comma, list_of_wffs(Z), assign(X,[Y|Z]) ) ) ).

comma --> ",".
turnstile --> "|-".

strip_of_spaces([],[]).

strip_of_spaces([H|T],T1) :-
    [H] = " ",
    strip_of_spaces(T,T1).

  strip_of_spaces([H|T],[H|T1]) :-
    strip_of_spaces(T,T1).
```

Exercise 5.7

```
                        /* Tree is an atom */

exp_tree(['A',Tree],Tree) :- !.

            /* Dominant connective is Binary Operator */

exp_tree(    [ _ ,[Left,Bin_Op,Right]],
             [Bin_Op,Left_1,Right_1] ) :-
    !,
    exp_tree(Left,Left_1),
    exp_tree(Right,Right_1).

                /* Dominant connective is Negation */

exp_tree(    [ _ ,['neg',Right]],
             ['neg',Right_1]) :-
    !,
    exp_tree(Right,Right_1).

                        /* brackets */

exp_tree([ _ ,['(',Tree,')']],Tree_1) :-
    exp_tree(Tree,Tree_1).
```

```
        /* further recursion eliminating non-terminals */

exp_tree([ _ ,Tree],Tree_1) :-
   exp_tree(Tree,Tree_1).

                                        /* */
```

Exercise 5.8

```
:- op(30,xfy,'<->').
:- op(28,xfy,'->').
:- op(26,xfy,'v').
:- op(24,xfy,'&').
:- op(20,fx,'-').

analyse(X,Y) :-
   atom(X),!, Y = X
   ;
   X =.. List,!,
   (    List = [-,Wff],
        analyse(Wff,Z),Y=[-,Z]
        ;
        List = [Op,Wff1,Wff2],
        analyse(Wff1,Z1),
        analyse(Wff2,Z2),
        Y = [Op,Z1,Z2]      ).
```

Exercise 6.2

```
add_to_list(Atom,[],[Atom]).

add_to_list(Atom,[Atom | Tail],[Atom| Tail]).

add_to_list(Atom,[Atom_1 | Tail],List) :-

    name(Atom,[N]),
    name(Atom_1,[N_1]),
    (     N < N_1, List = [Atom,Atom_1 | Tail]
          ;
          add_to_list(Atom,Tail,List_1),
          List = [Atom_1 | List_1] ).
```

Exercise 6.3

```
evaluate(Val_Tree,Value):-
   atom(Val_Tree),Value = Val_Tree,!
   ;
   Val_Tree = neg(Val_Tree_1),
   evaluate(Val_Tree_1,Val_1),
   (     Val_1 = true,
         Value = false
         ;
         Value = true    )
   ;
   Val_Tree = and(Val_Tree_1,Val_Tree_2),
   evaluate(Val_Tree_1,Val_1),
   evaluate(Val_Tree_2,Val_2),
   (     Val_1 = true,
         Val_2 = true,
         Value = true
         ;
         Value = false   )
   ;
   Val_Tree = or(Val_Tree_1,Val_Tree_2),
   evaluate(Val_Tree_1,Val_1),
   evaluate(Val_Tree_2,Val_2),
```

```
(       Val_1 = false,
        Val_2 = false,
        Value = false
        ;
        Value = true    )
;
Val_Tree = if(Val_Tree_1,Val_Tree_2),
evaluate(Val_Tree_1,Val_1),
evaluate(Val_Tree_2,Val_2),
(       Val_1 = true,
        Val_2 = false,
        Value = false
        ;
        Value = true    )
;
Val_Tree = iff(Val_Tree_1,Val_Tree_2),
evaluate(Val_Tree_1,Val_1),
evaluate(Val_Tree_2,Val_2),
(       Val_1 = Val_2,
        Value = true
        ;
        Value = false   ).
```

Exercise 7.1

<wff_in_cnf>	::=	<clause> | <clause> **&** <wff_in_cnf> | (<wff_in_cnf>)
<clause>	::=	<literal> | <literal> **v** <clause> | (<clause>)
<literal>	::=	<atom> | ¬<atom> | (<literal>)

Exercise 8.1

```
        /* CONJUNCTIVE NORMAL FORM for propositional
                            calculus */

:- consult('pc_wffs.3').

cnf(Wff,Cnf) :-
   wff(Tree,Wff,[]),
   stage_1(Tree,X),!,
   stage_2(X,Y),!,
   stage_3(Y,Cnf),!.

                          /* stage_1 */

stage_1(iff(L,R),and(or(neg(L1),R1),or(neg(R1),L1)) :-
   stage_1(L,L1),
   stage_1(R,R1).

stage_1(if(L,R),or(neg(L1),R1)) :-
   stage_1(L,L1),
   stage_1(R,R1).

stage_1(and(L,R),and(L1,R1)) :-
   stage_1(L,L1),
   stage_1(R,R1).

stage_1(or(L,R),or(L1,R1)) :-
   stage_1(L,L1),
   stage_1(R,R1).

stage_1(neg(R),neg(R1)) :-
   stage_1(R,R1).

stage_1(X,X).


stage_2(neg(neg(R)),R1) :-
   stage_2(R,R1).
```

```
stage_2(neg(and(L,R)),or(L1,R1))  :-
   stage_2(neg(L),L1),
   stage_2(neg(R),R1).

stage_2(neg(or(L,R)),and(L1,R1))  :-
   stage_2(neg(L),L1),
   stage_2(neg(R),R1).

stage_2(and(L,R),and(L1,R1))  :-
   stage_2(L,L1),
   stage_2(R,R1).

stage_2(or(L,R),or(L1,R1))  :-
   stage_2(L,L1),
   stage_2(R,R1).

stage_2(neg(R),neg(R1))  :-
   stage_2(R,R1).

stage_2(X,X).


stage_3(or(A,B),or(A,B))  :-
   is_clause(A),
   is_clause(B).

stage_3(or(A,and(B,C)),and(L1,R1))  :-
   stage_3(or(A,B),L1),
   stage_3(or(A,C),R1).

stage_3(or(and(B,C),A),and(L1,R1))  :-
   stage_3(or(B,A),L1),
   stage_3(or(C,A),R1).

stage_3(or(A,B),Z)  :-
   stage_3(A,X),
   stage_3(B,Y),
   stage_3(or(X,Y),Z).
```

```
stage_3(and(L,R),and(L1,R1))  :-
   stage_3(L,L1),
   stage_3(R,R1).

stage_3(X,X)  :-
   is_literal(X).


is_clause(X) :-
   is_literal(X),!
   ;
   X = or(A,B),
   is_clause(A),
   !,
   is_clause(B).

is_literal(X)  :-
   is_positive_literal(X)
   ;
   is_negative_literal(X).

is_negative_literal(neg(L))  :-
   is_positive_literal(L).


is_positive_literal(PL)  :-
   atom(PL),
   name(PL,[P]),
   P > 96,!,
   not(P = 118),!,
   P < 123.
```

Exercise 8.2

```
:-  consult('cnf.2').
:-  consult('write_wff').
```

```
write_cnf(String) :-
    write_string(String),
    cnf(String,Cnf),
    nl,nl,
    write_wff(Cnf),
    nl.


write_string(String) :-
    String = [],!
    ;
    String = [Head | Tail],
    put(Head),
    write_string(Tail).
```

Exercise 8.3

```
make_clause_set(and(X,Y),Z) :-
    make_clause_set(X,X1),
    make_clause_set(Y,Y1),
    append(X1,Y1,Z).

make_clause_set(X,[Y]):-
    make_clause(X,Y).

make_clause(or(X,Y),Z) :-
    make_clause(X,Z1),
    make_clause(Y,Z2),
    union(Z1,Z2,Z).

make_clause(neg(Atom),[neg(Atom)]).
make_clause(Atom,[Atom]).

union([],X,X).
```

```
union([H | T], X, Y) :-
   union(T,X,Y1),
   (     occurs_in(H,Y1),
         Y = Y1
         ;
         Y = [H | Y1]    ).
```

Exercise 9.1

(1)

```
Prem   (1)  ¬(Pv¬P)
Prem   (2)  P
2      (3)  Pv¬P                    2vI
1,2    (4)  (Pv¬P)&¬(Pv¬P)          1,3&I
1      (5)  ¬P                      2,4¬I
1      (6)  Pv¬P                    5vI
1      (7)  (Pv¬P)&¬(Pv¬P)          1,6&I
       (8)  ¬¬(Pv¬P)                1,7¬I
       (9)  Pv¬P                    8¬E
```

(2)

```
Prem   (1)  ¬(¬Pv¬Q)
Prem   (2)  ¬P
2      (3)  ¬Pv¬Q                   2vI
1,2    (4)  ¬(¬Pv¬Q)&(¬Pv¬Q)        1,3&I
1      (5)  ¬¬P                     2,4¬I
1      (6)  P                       5¬E
Prem   (7)  ¬Q
7      (8)  ¬Pv¬Q                   7vI
1,7    (9)  ¬(¬Pv¬Q)&(¬Pv¬Q)        1,8&I
1      (10) ¬¬Q                     7,9¬I
1      (11) Q                       10¬E
1      (12) P&Q                     6,11&I
```

(5)

Prem	(1)	¬PvQ	
Prem	(2)	¬P	
Prem	(3)	P	
2,3	(4)	P&¬P	2,3&I
Prem	(5)	¬Q	
2,3,5	(6)	(P&¬P)&¬Q	4,5&I
2,3,5	(7)	(P&¬P)	6&E
2,3	(8)	¬¬Q	5,7¬I
2,3	(9)	Q	8¬E
2	(10)	P->Q	3,9->I
Prem	(11)	Q	
3,11	(12)	Q&P	3,11&I
3,11	(13)	Q	12&E
11	(14)	P->Q	3,13->I
1	(15)	P->Q	1,2,10,11,14vE

(10)

Prem	(1)	P&(QvR)	
1	(2)	P	1&E
1	(3)	QvR	1&E
Prem	(4)	Q	
1,4	(5)	P&Q	2,4&I
1,4	(6)	(P&Q)v(P&R)	5vI
Prem	(7)	R	
1,7	(8)	P&R	2,7&I
1,7	(9)	(P&Q)v(P&R)	8vI
1	(10)	(P&Q)v(P&R)	3,4,6,7,9vE

<ground_term> ::= <constant> |
<arbitary_name> |
<function_symbol> (<ground_term_list>)

<ground_term_list> ::= <ground_term> |
<ground_term> , <ground_term_list>

Exercise 11.1

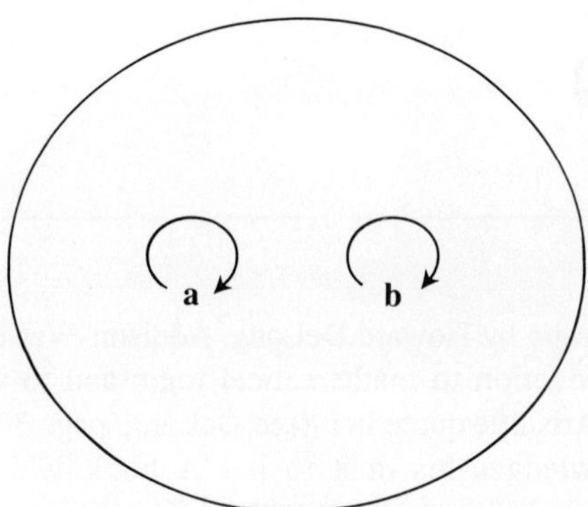

Suppose 'P(x,y)' is true in I iff x = y. Therefore the antecedent of $(\forall y)(\exists x)P(x,y) \text{ -> } (\exists x)(\forall y)P(x,y)$ is true in I. However, the consequent is false in I since there is no c in D such that <x,c> is a member of *P* in I. So the wff is not true in every I and so is not a logical truth[1].

Exercise 13.4(2)

Prem	**(1)**	$(\forall x)(\exists y)\neg(x=y)$	
1	**(2)**	$(\exists y)\neg(x=y)$	**1∀E**
1	**(3)**	$(\exists x)(\exists y)\neg(x=y)$	**2∃I**
	(4)	$(\forall x)(\exists y)\neg(x=y) \text{->} (\exists x)(\exists y)\neg(x=y)$	**1,3->I**

---

1 Philosophers will tell you that confusing

$$(\forall y)(\exists x)P(x,y) \Rightarrow (\exists x)(\forall y)P(x,y)$$

and

$$(\exists x)(\forall y)P(x,y) \Rightarrow (\forall y)(\exists x)P(x,y)$$

is the source of some of the commonest fallacies in metaphysics.

Further reading

A Profile of Mathematical Logic by Howard DeLong, Addison-Wesley (1970) is an excellent introduction to mathematical logic and to its cultural history. I found my Aristotle quote in it (see DeLong, page 34). Douglas Hofstadter acknowledges his debt to it ("A book which influenced me greatly") in his brilliant and eccentric *Godel, Escher, Bach: An Eternal Golden Braid*, Harvester Press (1979).

There are plenty of books on Prolog, but for my money the following texts stand out from the rest. *Programming in Prolog* by WF Clocksin and CS Mellish, Springer-Verlag, in various editions from 1981, was the first in the field, and anyone seriously interested in Prolog should work through it.

For an elegant treatment of logic programming see Christopher Hogger's *Introduction to Logic Programming*, Academic Press, London (1984). *The Art of Prolog* by Leon Sterling and Ehud Shapiro, MIT Press (1986), is, in my opinion, a beautiful book. It covers the foundations of Prolog as a logic programming language, as well as its use in practical programming problems.

For lots of examples of code, and for a thorough account of the peculiarities of Prolog in practice, see *Prolog: A Relational language and Its Applications* by John Malpas, Prentice-Hall International (1987). A good compendium of Prolog programming techniques can be found in *Logic Programming and Knowledge Engineering* by Tore Amble, Addison-Wesley (1987).

The standard source paper on the Prolog Grammar Rules is 'Definite clause grammars for language analysis - a survey of the formalism and a comparison with augmented transition networks', FCN Pereira and DHD Warren, Artificial Intelligence **13** (180) 231-278. The standard text-book account is to be found in Chapter 9 of *Programming in Prolog*, by WF Clocksin and CS Mellish. An extended treatment of parsing with Prolog can be found in the excellent *Prolog and Natural Language Analysis*, by FCN Pereira and SM Shieber Center for the

Study of Language and Information.

Books on natural deduction in a style similar to that adopted here include *Beginning Logic* by EJ Lemmon, Van Nostrand (1984), and *Logic An Introductory Course* by WH Newton-Smith, Routlege and Kegan Paul (1985). Our treatment of natural deduction is based closely on that of the latter.

Program proving for imperative languages is handled very well in *Program Construction and Verification*, by Roland C Backhouse, Prentice-Hall (1986).

For the theory of logic programming see *Introduction to Logic Programming* by CJ Hogger, Academic Press (1984), and *Foundations of Logic Programming* by JW Lloyd, Springer-Verlag (1984), *The Computer Modelling of Mathematical Reasoning* by Alan Bundy, Academic Press (1983) and *Logic for Problem Solving* by Robert Kowalski, North-Holland (1979). The last of these is one of the classics of the subject.

Finally, *Computing with Logic* by David Maier and David S Warren, Benjamin, 1988, has the best account I have seen of the logic of Prolog together with a discussion of the problems of implementing Prolog via interpreters and compilers written in imperative programming languages.

References

Amble, Tore (1987). *Logic programming and knowledge engineering*. Addison-Wesley.

Backhouse, Roland C (1986). *Program construction and verification*. Prentice-Hall.

Bundy, Alan (1983). *The computer modelling of mathematical reasoning*. by Academic Press.

Berlioux, P and Bizard, P (1986). *Algorithms: the construction, proof, and analysis of programs*. John Wiley.

Clocksin, WF and Mellish, CS (1981). *Programming in Prolog*. Springer-Verlag.

Dahl, O-J, Dijkstra, EW and Hoare, CAR (1972). *Structured programming,* Academic Press.

deLong, Howard (1970). *A profile of mathematical logic*. Addison-Wesley.

Dromey, RG (1982). *How to solve it by computer*. Prentice-Hall.

Gentzen, G (1969). *The collected papers of Gerhard Gentzen*. (ed ME Szabo). North-Holland.

Goldschlager, L and Lister, A (1982). *Computer science: a modern introduction*. Prentice-Hall.

Gries, D (ed) (1978). *Programming methodology*. Springer-Verlag.

Hilbert, David and Ackermann, Wilhelm (1950). *Principles of mathematical logic.* Chelsea Publishing Company. A translation of their *Grundzuge der theorischen Logik* of 1928.

Hofstadter, Douglas (1979). *Godel, Escher, Bach: an eternal golden braid.* Harvester Press.

Hogger, CJ (1984) *Introduction to logic programming*. Academic Press.

Hunter, Geoffrey (1971). *Metalogic*. University of California Press.

Kowalski, Robert (1970). *Logic for problem solving*. Elsevier North-Holland.

Lemmon, EJ (1984). *Beginning logic*. Van Nostrand.

LLoyd, JW (1984). *Foundations of logic programming*. Springer-Verlag.

Maier, David and and Warren, David S (1988). *Computing with logic.* Benjamin.

Malpas, John (1987). *Prolog: a relational language and its applications.* Prentice-Hall International.

Manna, Z and Waldinger, R (1985). *Logical basis for computer programming*. Addison-Wesley.

Newton-Smith, WH (1985). *Logic: a first course,* Routledge and Kegan Paul.

Pereira, FCN and Warren, DHD (1980) 'Definite clause grammars for language analysis - a survey of the formalism and a comparison with augmented transition networks.' *Artificial Intelligence* 13 (180) 231-278.

Pereira, FCN and Shieber, SM (1987) *Prolog and natural language analysis*. CSLI Lecture Notes Number 10, Center for the Study of Language and Information.

Prawitz, Dag (1965). *Natural deduction: a proof-theoretical study.* Almqvist and Wiksell.

Quine, WVO (1952). *Methods of logic.*Routledge and Kegan Paul.

Rucker, Rudy (1988). *Mind tools: the mathematics of information.* Penguin Books.

Russell, Bertrand and Whitehead, Alfred North (1910-13). *Principia Mathematica* . Cambridge University Press.

Sterling, Leon and Shapiro, Ehud (1986) *The art of Prolog.* MIT Press.

Stoy, J (1977). *Denational semantics.* MIT Press.

Wirth, Niklaus (1976). *Algorithms + Data Structures = Programs.* Prentice-Hall.

Index